MW00473469

Concise Guide to
Laminitis in the Horse

Also by David W. Ramey, DVM

The Anatomy of a Horse
Concise Guide to Arthritis in the Horse
Concise Guide to Colic in the Horse
Concise Guide to Medications and Supplements for the Horse
Concise Guide to Navicular Syndrome in the Horse
Concise Guide to Nutrition in the Horse
Concise Guide to Tendon and Ligament Injuries in the Horse
A Consumer's Guide to Alternative Therapies in the Horse
Horsefeathers: Fact vs. Myths about Your Horse's Health

Concise Guide to
Laminitis in the Horse

David W. Ramey, DVM

Trafalgar Square Publishing
North Pomfret, Vermont

First published in 2003 by
Trafalgar Square Publishing
North Pomfret, Vermont 05053

Printed in Canada

Copyright © 2003 by David W. Ramey, DVM

All rights reserved. No part of this book may be reproduced, by any means, without written permission of the publisher, except by a reviewer quoting brief excerpts for a review in a magazine, newspaper, or web site.

Disclaimer
This book is not to be used in place of veterinary care and expertise. The author and publisher shall have neither liability nor responsibility to any person or entity with respect to any loss or damage caused or alleged to be caused directly or indirectly by the information contained in this book. While the book is as accurate as the author can make it, there may be errors, omissions, and inaccuracies.

Library of Congress Cataloging-in-Publication Data

Ramey, David W.
    Concise guide to laminitis in the horse / David W. Ramey.
        p. cm.
    ISBN 1-57076-237-6 (pbk.)
    1. Lameness in horses. I. Title.

    SF959.L25 R35 2003
    636.1'089758--dc21

                                                    2002152218

Cover and book design by Lynne Walker Design Studio
www.lynnewalker.com
Typeface: Leawood and Sabon

10 9 8 7 6 5 4 3 2 1

CONTENTS

ACKNOWLEDGMENTS

Once again, I am indebted to a good number of people who helped me with suggestions, information, and clarity. Linda Rarey has read through every one of these books prior to publication and she always finds something to make them better. Valeri Devaney, now taking care of horses on her own, somehow found time between classes and rounds to add her invaluable insights; thanks also to Valeri's Mom, who got conned into helping, too. Robert Amaral brings his expertise from the field of medical illustration to illuminate the pages, and the book is much the better for it.

A number of prominent and very smart veterinarians were gracious enough to review some very important information and add the weight of their expertise to this book. Dr. Bill Moyer at Texas A&M has worked with laminitic horses for 30 plus years—I hope he'll figure it out eventually. Dr. Chris Pollitt has done some wonderful work "Down Under" that brings us ever closer to solving this terrible problem. Ditto Dr. Philip Johnson at the University of Missouri, who is especially interested in solving the perplexing problem of why some fat (horizontally challenged) horses tend to get laminitis. Dr. Nat Messer, a former professor of the author, and teaching students still, was kind enough to thoroughly review the material on medical treatments. Dr. Steve O'Grady combines the skills of a terrific veterinarian and talented farrier—I just hope that his back holds out. Finally, Dr. Chris King provided some wonderful insights from her book *Preventing Laminitis*. I'm very lucky to have had such wonderful, expert, and enthusiastic help.

Finally, even though it's getting harder and harder to find time to fit writing in between trips to the driving range and the tennis court, it's always a pleasure. Jackson and Aidan, I hope that you can find the time to be both firemen and horse doctors.

INTRODUCTION

Laminitis is a potentially devastating disease of the horse's feet. While many horses can recover from an episode of laminitis uneventfully, the disease can also end a horse's career, or even its life. It's a bane to horse owners, farriers, and veterinarians alike (not to mention the horses).

Fortunately, intensive investigations have shed light on what's underlying this troubling condition. While a sure cure is still just a dream, many serious cases are now being saved that might not have been in years past. And, good strategies for prevention exist in many cases. While it's likely that laminitis will always be a concern, good management and good care (and some good luck) give you the best chance to overcome this frustrating disease.

The book begins with an overview of laminitis, both in historical terms and in the sense of looking inside the horse's foot (which is where all of the action occurs). From there, it moves to the most current thought about what happens inside the foot of the horse. Obviously, to effectively treat a problem, you have to know that a problem exists! That's why Chapter Three is devoted to helping you understand how a diagnosis of laminitis is made. Hand-in-hand with making a diagnosis is being aware of the underlying causes. Therefore, the fourth chapter discusses the seemingly innumerable and apparently unrelated things that are associated with the development of the problem. Once you've identified that there is a problem, you'll want to treat it; thus the next three chapters discuss medical and surgical care, and perhaps most importantly, care of the foot of the affected horse. It's worth your effort to try to prevent laminitis rather than treat it and as such, you'll certainly appreciate the next-to-last chapter; laminitis is a problem that you and your horse are better off without. Finally, some important counseling information is offered in the last chapter so that you'll be able to maintain a reasonable and rational approach to the problem.

Laminitis is a disease that appears to occur largely because people have domesticated the horse. By taking it out of its natural free-ranging situation, the horse has become overfed and under-exercised. It no longer follows its normal feeding or activity patterns—its situation has been changed to accommodate the needs of its owner. The price for this convenience seems to be an increase in a variety of medical problems, including laminitis.

The information that you find in this book will help you formulate a rational approach to the prevention and treatment of laminitis. Please read the book carefully; discuss it with your friends, your trainer, and veterinarian. Ask your veterinarian if you have any questions about your horse's health because laminitis is a serious problem. The more you know, the better off you and your horse are going to be.

1

What is Laminitis?

The horse has been domesticated for centuries. Without a doubt, laminitis has been a problem for just as long. Over the years, there have been countless attempts at fixing the problem, including various medicines, surgical approaches, and innumerable shoeing techniques. These approaches have all had their vocal proponents and "successes"—and they have all had complete failures. It's only been a relatively short time that good scientific investigations have been directed at the problem; as a result, knowledge about the condition has increased dramatically. With all of the eager and well-meaning attempts to solve this troubling problem, there's a witches brew of scientific data, clinical impressions, and old myths that you'll have to sort through when trying to understand what's going on in a horse with laminitis.

The word roots that give meaning to the term laminitis are really quite simple. And, when you know something about the anatomy of the foot, it's also quite descriptive. Any time that you see the suffix "-itis" added on to a word, it means that something is inflamed. In the case of this disease, that something is the laminae of the hoof, which are the connections that bind the bone of the hoof to the hoof itself (you'll read all about them in a bit). Still, even though inflammation does occur in the foot of the horse with laminitis, there's a good bit more to the condition than just treating and attempting to relieve inflammation. In fact, laminitis is a complicated condition that may progress to involve virtually all of the anatomical structures of the horse's foot.

HISTORICAL ASPECTS

Diseases predate people's understanding of them. Nevertheless, people have apparently always wanted to put a name on disease conditions, probably just so that they could identify them and call them *something*. As a result of this eternal attempt to put a name on things that they didn't understand, people are still using terms like malaria (the "mal" means bad and the "aria" refers to the air) or influenza (implying that the disease comes from evil influences). In the case of laminitis, people didn't always know about the laminae of the hoof and they didn't always understand the process of inflammation. So, over history, there have been all sorts of terms used to describe the condition. For example, the ancient Greeks referred to laminitis as "gout" or "barley disease." Of course, these terms aren't used anymore, but some old words do seem to have a way of sticking around. In the case of laminitis, there is a very old word that's still in use. That word would be "founder."

The word "founder" apparently first showed up in Geoffrey Chaucer's writings from the thirteenth century. People who study the origins of words think that "founder" most likely comes from an Old French verb that meant "to sink." And, as such, it's really not such a bad term. In fact, in some horses—those in which the bone "sinks" into ground (and out of the hoof)—it sometimes describes the situation pretty well.

Some people have tried to make a distinction between the terms *founder* and *laminitis*. For example, for a while, laminitis was the veterinarian's term and founder was the term used by everyone else. Others have tried to say that a foundered horse is one in which there is chronic lameness, accompanied by changes in the anatomical structures of the foot. Such distinctions really aren't particularly useful and they can be quite confusing. In fact, they're simply all just different degrees of the same thing.

To begin to understand what's going on inside the hoof of a horse with laminitis, you have to know just a little bit about the structures that become involved with the disease process. Once you know a few basics about the relevant anatomy, you're on your way to understanding why this can be such a difficult and complicated disease.

A BRIEF ANATOMY OF THE HORSE'S FOOT
While the anatomy of all of the structures that are hidden inside the horse's hoof is certainly quite complex, the anatomy that's relevant for an understanding of laminitis, and for understanding the rationale for its various treatments, is really not all that complicated. In fact, it comes down to a few key structures (Figure 1).

FIGURE 1

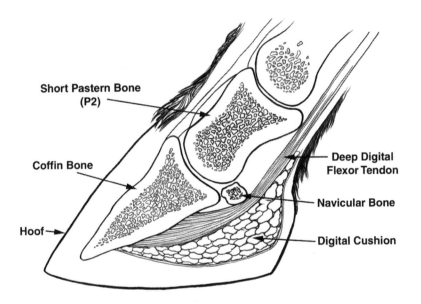

Short Pastern Bone (P2)

Coffin Bone

Hoof

Deep Digital Flexor Tendon

Navicular Bone

Digital Cushion

Some important anatomy of the horse's hoof.

The Hoof

The horse's hoof is a marvelous structure that's strong, flexible, renewable, and invaluable. It's a remarkable configuration that is uniquely strong and shaped so as to carry the great load of the weight of the horse. It grows down from the coronary band (located where the hair meets the hoof) and totally replaces itself every 4–12 months. At the heels, where the hoof is shorter, it takes less time to replace itself than at the toe, where it is longer.

Hoof grows down from the coronary band in small little tubes. These tubes are oriented vertically, roughly perpendicular to the ground. The tubes grow down in a solid sheet, like a wall. In fact, the portion of the hoof that you can easily see when the horse's foot is on the ground is called the *hoof wall*.

Ordinarily, you can get some indication that the wall is made up of little tubes by two kinds of lines that you normally see on the hoof wall. If you look closely, you'll see regular fine lines that run perpendicular to the ground, called the *tubules*. Then, there are more variable, irregular lines that parallel the coronary band. These parallel lines are the rows of the cells that lock the tubules together.

In abnormal hooves, you'll sometimes see larger, irregular lines around the hoof wall that run at right angles to the tubules, or roughly parallel to the coronary band. These lines appear as "rings" around the hoof. The rings may be normal. For example, normal rings in the hoof wall may occur as a result of such things as changes in the seasons or with changes in nutrition. And, rings in the hoof wall can also be abnormal, occurring after such things as fevers or infection. But in all cases, rings reflect the fact that something happened to the horse in the past, and, in particular, hoof rings may also be seen after episodes of laminitis (so keep an eye open for such things in a horse that you may be looking to buy).

Laminitis affects the strength of the hoof. Therefore, trying to protect and restore hoof strength becomes fundamental when trying to treat horses with laminitis. Of course, since 1) People like to try to fix problems; 2) in laminitis, the problem is in the hoof; 3) the hoof is right there and; 4) it's not particularly difficult to pare and shape a hoof, a great deal of attention is usually given to the hoof of a horse affected with the laminitis. However, although many different approaches to treating the hoof have been tried, no single approach has been successful. Unfortunately, this means there are many different opinions. To help you sort through them, there's a whole chapter on the subject of foot care and laminitis later in this book.

The Coffin Bone

Inside the hoof sits the coffin bone. It's called the coffin bone because the bone appears as if, according to an eighteenth century source, it is inside a coffin (morbid, but true). The main function of bone—all bone—is to serve as the structural framework for the horse's body. Bones support the massive weight of the horse and also serve as attachments for all sorts of other tissues.

So it is for the coffin bone, which is particularly critical because all of the weight of the horse ultimately comes to bear on it. Accordingly, it's subject to tremendous stresses. Although it may not look like it, bone is a living and very active tissue, and when stress is applied to it, the bone changes. Laminitis causes additional stresses; thus, laminitis may be associated with any number of changes in the coffin bone.

The Laminae

The hoof and the coffin bone are connected. These connections, called the laminae, are the only things that keep the weight of the horse from driving the coffin bone right on down into the ground (Figure 2). The laminae link up like Velcro®; some of them arise from the inner part of the hoof wall and others come from the living tissue that's attached to the surface of the coffin bone. The laminae are the key areas affected in laminitis.

These attachments between bone and hoof are simply thin little layers (in fact, the word "lamina" is Latin for layer). They interlock with each other in a manner not unlike the paper that's around a popular peanut-butter-and-chocolate candy cup. There are hundreds of them inside each foot. A thin sheet of tissue known as the basement membrane keeps these layers attached. The basement membrane is the glue that holds the laminae of the hoof to the laminae of the coffin bone. When laminitis occurs, it's this basement membrane that seems to be the target of the initial damage. When the basement membrane starts to fall apart, the connections between the hoof and the coffin bone may begin to give way, with some potentially devastating consequences for the poor horse to which those connections are attached.

FIGURE 2

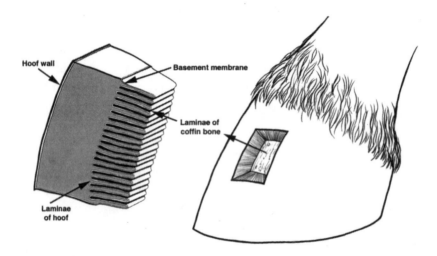

The laminae attach the hoof to the underlying structures. The basement membrane is where much of the action in laminitis appears to occur.

The Deep Digital Flexor Tendon

A tendon connects a muscle to a bone. In the particular instance of laminitis, the tendon is the deep digital flexor and the bone is the coffin bone. The muscles that pull on the coffin bone via the deep digital flexor tendon start above the horse's knee (carpus) or hock (tarsus) in the front or hind limb, respectively; as they continue down toward the ground, the muscles narrow into tendon at about the level of those major joints. From there, the deep digital flexor tendon drops down the back of the horse's cannon bone and attaches to the bottom of the coffin bone, underneath the toe (see Figure 1). When the muscle contracts, which is the only thing that muscle can do, it allows the horse to pull its foot up and back during its stride.

In the normal horse, this muscle-tendon system exists under tension, much like the string on an archer's bow. So, in addition to making the movement of the horse's leg possible, the system acts like a supporting cable for the horse's leg. The horse's weight wants to push everything down into the ground; the muscle-tendon system resists the downward force on the leg that comes from the weight of the horse above it. Unfortunately, the normal tension that exists in the deep digital flexor tendon can become something of a problem in laminitis, especially in the front leg, where the horse bears most of its considerable weight.

The Blood Vessels

Finally, just a quick word about the blood supply to the foot. Of course, blood is critical for any living tissue. Blood, containing oxygen and other nutrients, is carried to the foot in arteries. It's taken away from the foot in veins. There are two big arteries that run down the back of the horse's pastern; right next to them are two big veins. In between arteries and veins are very small blood vessels called capillaries. Capillaries get the blood into every nook and cranny of the foot.

Having veins, arteries, and capillaries does not distinguish the horse's foot from any other tissue in the horse's body. However, in the foot, the circulatory structures have some very special and unique functions. For one, the blood itself helps to cushion the structures of the hoof. The blood in the vessels appears to act very much like the water in a waterbed, cushioning the horse on a liquid foundation. Damage to these little vessels can affect the ability of the horse to comfortably support itself.

But there's another really interesting thing about the blood supply to the horse's foot. In addition to the capillaries that run throughout the foot, there are other direct connections between the arteries and the veins. These connections, called shunts, are found at about the level of the hairline, at the top of the foot. The shunts allow the circulating blood to bypass the foot.

Of course, normally, blood doesn't bypass the foot, but the shunts do serve a useful purpose in certain circumstances. It is thought that the shunts are there to help the horse regulate the temperature of its feet. They allow the horse to adjust the flow of the warm blood to the hoof, allowing for more or less blood as environmental temperatures dictate. Indeed, some interesting observations about these shunts were made on the temperature of the feet of normal horses in a climate-controlled laboratory in Australia. Investigators there found that hoof temperature is normally the same as the temperature of the surrounding environment, at least for most of the time. That is, at whatever temperature a controlled environment is set, between 32–86 degrees Fahrenheit (0–30 degrees Celsius), the temperature of the surface of the hoof matches it most of the time. However, when the environment is cold, peaks and valley in the temperature inside the hoof occur. In cold temperature, the hoof gets cold, but then the temperature rapidly changes and can reach as high as 86 degrees F (30 degrees C). The researchers attributed this quick change in temperature to the shunts opening up and letting more warm blood in.

This sort of thing is seen in the real world, as well as the laboratory. For example, in horses tested in Alaska, when the outside temperature was as cold as 80 degrees F (30 C) below freezing, hoof temperature was 1–2 degrees above freezing but never below. And, the temperature of the hoof periodically rose to around 86 degrees F (30 C) for an hour or two every day. The reverse probably occurs in very hot places: via the shunts, the hooves are kept at blood temperature in the face of air temperatures that will fry eggs. Otherwise stated, if it's really hot outside, the shunts might close to allow the blood to bypass the foot—if it's really cold outside, they might not be used at all. It's really quite a marvelous system—and it's automatic!

The shunts can become very important in the horse with laminitis. Under the abnormal conditions associated with the disease, the shunts appear to allow blood to bypass the feet (which is most likely a bad thing, although no one knows for sure). In any case, it's fair to say that the unique and special features that characterize the blood circulation to the horse's foot also make it a target for the numerous problems that can occur in laminitis.

Now that you've got a basic understanding of the relevant anatomy of the structures of the horse's hoof that become involved in laminitis, the next step is to try to understand what goes wrong with this intricate and elegant system. While the exact events leading up to the development of laminitis still aren't completely understood, investigators are getting closer and closer to defining the underlying problem.

2

What Happens?

with Chris Pollitt, DVM

At its simplest level, laminitis is the failure of the attachments—the *laminae*—between the coffin bone of the foot and the inner surface of the hoof wall. If the bone of the hoof isn't properly attached inside the hoof, the weight of the horse and the enormous forces associated with normal movement can drive the bone right down into the ground. In the process, virtually all of the important tissues inside the hoof can potentially be damaged. This damage is in a large part responsible for the unrelenting pain and lameness that is often associated with the condition.

But it's the earliest microscopic changes that occur in the development of laminitis that are of most interest to those working to solve the problem. If these early underlying changes can be illuminated and understood, it's possible that newer and better ways to treat, and perhaps more importantly, to prevent, the disease might be developed. That said, the details associated with the failure of the laminae inside the hoof are just starting to be worked out. Currently, there are three main theories of what happens, and, to some extent, they overlap. Understanding what's happening inside of the foot of the horse with laminitis can go a long way toward explaining why the problem is often so difficult to manage.

THREE PHASES OF LAMINITIS

The clinical disease of laminitis is commonly broken dow
three separate phases—developmental, acute, and chro
reflects the fact that different things may be occurrin
times. It also implies that different considerations m

depending on which phase of the disease that is being treated. Sure, it's all part of the same disease, but, as the old saying goes, the devil is in the details.

The first phase of laminitis has been termed the *developmental* phase. During this phase, the trouble starts; the separation of the laminae attaching the coffin bone and the hoof begins. Unfortunately, you may not know that this phase is occurring because horses in the developmental phase of laminitis generally won't show any obvious clinical signs of the disease, such as pain. The developmental phase may be as short as 8–12 hours in the cases of horses exposed to toxic substances like black walnut shavings or 30–40 hours in the cases of horses that have eaten too much grain (more on this later).

During the developmental phase of laminitis, and before you start to see clinical signs of pain, horses or ponies usually experience problems with any one of a number of body systems. The initial event that ultimately results in laminitis may start in the intestines, the respiratory system, the reproductive system, the endocrine system, or just about anywhere else. Whatever the initial event and in whichever system it occurs, it may start a series of events that ultimately expose the connections in the hoof to certain "factors" that lead to separation and/or distortion of the intricate structures of the foot. Unfortunately, no one yet knows the exact nature of the "factors" that reach the foot via the circulation (but people are looking).

As the disease progresses, the developmental phase of laminitis becomes the *acute* phase. The acute phase lasts from the time the first clinical signs of foot pain become evident until such time as changes in the foot can be seen on X rays, or, if such changes don't occur—and, happily, they don't always occur—the horse gets better. It's generally felt that prompt and aggressive treatment, as early as possible in the acute phase of the disease, gives you the best chance of success in treating laminitis. This makes some sense of course; the sooner you get on with the job of putting out a fire, the less damage is likely to be done to the house.

If the initial disease process doesn't kill the horse, or if the horse doesn't make a complete recovery, horses may enter the third, *chronic* phase of laminitis. This phase can last indefinitely. Clinical signs of chronic laminitis may range from a persistent, mild lameness, to severe, unrelenting foot pain; from no noticeable abnormalities after a successful recovery, to permanent hoof deformities; to loss of the whole hoof. Many horses affected with laminitis recover; some never return to normal; some even die.

THE ROOT OF THE PROBLEM

The end result of laminitis is a spectacular *disintegration of the laminae* that attach the coffin bone to the hoof. This disintegration starts during the developmental phase of laminitis. You will not even know that it is happening. The breakdown of the hoof architecture that may result from laminitis can turn a system that is normally remarkably resilient and trouble-free, into one that is virtually useless, in a relatively short period of time. Exactly what happens is still a puzzle. Fortunately, however, some of the parts of the puzzle have begun to come together. People have minutely dissected the anatomy, exhaustively discussed the physiology, examined all of the structures under a microscope, and conducted all sorts of experiments in an effort to understand what's happening in the earliest stages of the disease. Still, even as information accumulates, exactly how, and in what sequence, the events leading up to laminitis occur has yet to be figured out.

As such, there are currently three prevailing theories that attempt to describe what's going on in the earliest stages of laminitis. The first, and oldest theory, suggests that laminitis results from changes in the circulation of blood to the foot. The second theory suggests that laminitis is a result of trauma, which directly damages the structures of the hoof. The third and newest theory, suggests that laminitis is a result of out-of-control enzyme activity in the hoof. There is some overlap among the theories. There are observations that support each of them, although none of them appears to completely explain what's going on. All of them need to be understood in order to talk about how to best try to help a horse with laminitis.

THE CIRCULATION THEORY

In the *circulation theory,* all of the problems that occur as a result of laminitis can ultimately be traced back to an inability of the horse's body to continue normal blood flow to the foot. That is, during the developmental phase of laminitis, it's proposed that the foot loses some part of its blood supply. According to the theory, increased pressure inside the hoof, which is said to occur as a result of the "factors" that trigger the disease (whatever they are), seriously interferes with the flow of blood to the foot. Of course, normal blood supply is absolutely essential for any tissue, and losing blood supply is a terrible thing. When the blood supply is removed from any tissue, it becomes painful and eventually dies. Once tissue, in this case, the laminae of the feet, dies, you end up with permanent changes and irreversible damage.

The circulation theory has several observations to support it. Early in the onset of laminitis, it has been shown that the temperature of the hoof wall decreases. This suggests that there's a reduced blood supply to the foot. Reduced blood supply would ordinarily be associated with cell and tissue death and that's seen in laminitis. It also might explain why horses don't experience pain early in the disease, since tissue that has reduced blood supply also has reduced sensation. Eventually, as the circulation tries to recover (and it does try to recover) the pain starts. Some people even feel that there are additional problems caused by the horse's body's efforts to try to get the circulation going again (so-called "reperfusion injury"), but it's not at all clear how important this is and it won't be discussed here.

Of course, the circulation theory isn't the whole answer, either. There are some problems with it. For example, the sorts of things that you'd expect to see in tissue that's lost its blood supply, such as blood clots and swelling, aren't seen in microscopic sections taken from the hooves of horses in the early stages of laminitis.

Still, if the circulation to your horse's foot is impaired, it implies several things. First of all, it means that a whole lot of damage to the foot can occur before anyone would ever be aware of it. Second, it implies that the worse the impairment to the circulation, the greater the damage. Third, it certainly seems to suggest that, as with just about any medical condition, the sooner you can start trying to get things under control, the less damage there might be, and the better off things might end up for the horse.

From the standpoint of treatment, if there's reduced blood supply to the foot, it suggests that you ought to try to do things to keep the blood flowing there. Predictably, that's exactly what many of the therapies for laminitis try to do. It also suggests that you need to be awfully careful in handling a horse with laminitis; if the tissue is damaged because there's not enough blood flowing, you don't want to risk making the situation worse by trailering, exercising, or even shoeing the horse when the important structures of the hoof first start going "south."

However, even though there's little doubt that blood flow to the foot is affected in laminitis, it's not known if this is the primary cause of the problem, if it's just one of many things that contributes to the problem, or if it's simply a change that occurs as a result of other things that happen with laminitis. It's a chicken-and-egg sort of controversy. However, those are the sorts of little, but very important, details that still need to be worked out in order to come up with a solution for the problem of laminitis.

THE TRAUMA THEORY

Traumatic theories of laminitis suggest that many of the problems that occur due to laminitis are simply caused by injury to the tissues of the hoof. Such things as *road founder*—the laminitis that shows up after heavy exercise—or the laminitis that can be seen in one leg when a horse bears all of its weight on it, such as happens after a painful surgery in the opposite leg, would appear to be due to some sort of mechanical overload of the foot. It's a pretty simple concept, really; the stress applied to the foot simply overwhelms the hoof's structures. The front or back half of the horse is simply too heavy to be held up by one leg only. Trauma can cause laminitis in any number of ways, from tearing of the laminae in the hoof, to blood-vessel damage from excessive weight bearing.

There haven't been any studies done on this, but it's generally assumed that *traumatic laminitis* is a different kettle of fish than the laminitis that occurs as a result of the factors that initiate laminitis in most other cases. Still, there are some common threads. For example, the stress of overloading the foot could cause inflammation, just as with the other theories. Trauma might also cause the blood vessels to spasm and start the foot down the road to the problems caused by reduced blood supply. Additional information about traumatic laminitis awaits future investigations.

THE ENZYME ACTIVATION THEORY

Recent research from Australia suggests that the root of the problems associated with laminitis is enzymatic changes that occur in the tissues of the foot that normally allow for growth of the hoof wall. The *enzyme theory* of laminitis challenges the *circulation theory* that has prevailed in veterinary medicine for some time.

Obviously, the horse's hooves, just like your fingernails, grow continually. For growth to occur, the hoof wall, which is made up of dead tissue, must be able to continually move over the stationary bone of the foot to which it is attached. The downward movement of the hoof wall (relative to the bone) as it grows is made possible by a continuous reshaping of the inner parts of the laminae, and several important enzymes assist in this.

Enzymes are simply substances (proteins) that help along chemical reactions; they do so without themselves getting used up in the process. In the case of the hoof, the important enzymes are called *proteinases* because they work on the proteins of the hoof wall. In the growing hoof, it's generally assumed that activity of these enzymes goes on pretty much all the time, due to the stresses and strains of normal horse movement and life. This enzyme activity also allows for normal hoof wall growth. Enough proteinases are made in the area when and where they are needed to allow for a little release, a little slip, of the microscopic connections between the cells of the bone of the hoof and the hoof wall. It's just enough of a move to allow for the hoof to slowly grow, while continuously maintaining the correct shape and alignment of the hoof structures. The proteinase enzymes most likely have other functions, as well. For example, from time to time, the structures of the hoof may be injured; this would require enzymes to be released to assist in repair and reconstruction of the damaged hoof. Of course, this sort of enzyme activity couldn't be allowed to go on willy-nilly, so under normal circumstances, the enzyme activity that occurs in the hoof is kept in check by specific enzyme inhibitors so that the whole process stays in balance. But as a result of it all, the growing hoof slowly moves past the stationary bone.

Problems arise when this elegant system gets out of whack. What appears to happen in laminitis is that production of the protein-dissolving proteinase enzymes increases. This sort of thing is seen in humans during the spread of malignant cancers. The switch from benign to malignant sees cancer cells not only making their own enzymes, but also activating enzymes in the tissues around the original growth. By dissolving proteins, these enzymes allow the cancer cells to cut their way out of the primary tissue, enter the blood vessels, and spread through the body at will. Proteinase enzyme levels increase in the foot of the horse with laminitis and it's very likely that some sort of factor triggers them. What's not yet known is what exactly causes the increase.

Regardless of the exact cause, when levels of proteinases start to increase in the hoof affected with laminitis, the intricate microscopic architecture of the hoof becomes distorted. The cells that make up the hoof-bone connections lose their normal shape and appear to slide over one another. The laminae that connect the hoof to the bone of the foot start to break down. The hoof starts to peel away from the underlying bone. If this goes on unchecked, there can be a failure of the entire hoof anatomy and the whole hoof can come off the horse's foot.

Also affected by this process are the blood vessels of the hoof. (Here's where things may start to overlap.) As the laminae between hoof and bone are destroyed, so are the little capillaries, which bring blood to those connections. As the capillaries get destroyed, blood that's trying to get to the tissues of the foot just can't get there—there's basically nowhere for it to go. The destruction of the capillaries does two things. First, it causes a tremendous increase in the resistance to the flow of blood. This has implications for the *diagnosis* of laminitis; it means that blood is going to have a hard time getting into the foot, and you may even be able to feel that such a thing is happening (more on that a bit later). Second, it causes blood to bypass the capillaries. This may have implications for the *treatment* of laminitis.

So, the key to treating laminitis under the enzyme activation scenario is to try to slow down the enzyme activity and to try to keep the factors that start the process from getting to the foot in the first place. No small challenge, of course, but there you go. And, it implies a quite different therapeutic approach, at least initially, in the *developmental phase* of laminitis. If you're trying to counteract the changes that occur during the enzyme theory, theoretically, in addition to trying to do things to interfere with the enzyme activity, you want to try to do things to *prevent* blood, which presumably carries that "something" that triggers the uncontrolled enzyme activity, from getting to the foot in the first place. Of course, this would be almost exactly the opposite of the stated goal of treating laminitis proposed by the circulation theorists, who suggest that you need to try to help keep the blood flowing to the foot. Such is the still-confusing world of laminitis.

AS TIME GOES BY

There's really not any controversy at all about what happens to the horse's foot as laminitis progresses through its phases. Of course, it doesn't always progress through the *acute* to the *chronic* phase—a good number of horses will get better (in some cases, even in spite of treatment). But, if laminitis does progress and become *chronic,* to a certain degree, the foot falls apart. Depending on how badly the foot falls apart, the horse may be able to recover, at least partially, or it may be damaged for the rest of its life. Some horses may have permanent changes in their feet, as shown on X rays, and yet still may be able to perform as useful athletes. Others may hurt so badly that the only humane thing to do is to put them to sleep.

Overall, it's possible to classify the problems that occur as time goes by into four categories.

1. Mechanical collapse of the foot and its supporting structures.

2. Problems with foot growth and metabolism.

3. Problems with the blood vessels.

4. Infection.

Mechanical Collapse of the Foot and Its Supporting Structures
The main underlying problem that's responsible for the mechanical collapse of the foot of the horse with laminitis is a failure of the attachments to the hoof wall—the laminae. Ultimately, as the laminae fail, they tear apart. Depending on how bad things are, how long it's been since the problem started, how much weight is placed on the foot and what damage has been done, the end result may be more or less severe.

Discussions about the structural failures of the foot that occur as a result of laminitis primarily focus on the coffin bone and its relationship to the hoof. As a result of laminitis, there may be movement of the coffin bone inside the hoof; that is, the bone may change its position. One of three things may happen to the relationship between coffin bone and hoof. In mild cases, or in cases that recover, it may not go anywhere at all (this is what you're hoping for). In more severe cases, the coffin bone may start to point downward and rotate relative to the hoof wall. In the most severe cases, virtually all of the laminar attachments between the coffin bone and the hoof fail and the bone "sinks" inside the hoof. But what do terms like "rotation" and "sinking" really mean?

If you look at an X-ray film of a side view of the horse's foot, the front of the coffin bone and the hoof wall are ordinarily parallel (Figure 3). You can draw lines and see for yourself. As laminitis progresses, and the bone moves in the foot, these lines change in relationship to each other. The line drawn along the front of the coffin bone may become steeper. If you connect the two lines until they intersect, you'll come up with an angle (remember geometry?). The degree of this angle, which is interpreted as the degree of rotation, can be used to give you some idea of how your horse is going to fare.

Rotation most likely occurs due to a combination of factors. The first is the pull of the deep flexor tendon on the tip of the coffin bone. This tendon tends to pull the tip of the coffin bone down toward the ground under normal circumstances. When the laminae are damaged, there's not much resistance to this pulling force and the bone may get pulled away from the hoof wall. Further adding to the problem is the fact that the limb with laminitis may be very painful, and this tends to make the horse reluctant to use the leg. A failure to use a limb can result in it being shortened due to contraction of the tendon(s). Adding to the tendency to rotate is the fact that the heel and toe of the horse with laminitis tend to grow at unequal rates: The heel of the horse with laminitis tends to lengthen, which causes the joint of the foot to flex (bend backward), and the toe to tend to point down. Finally, the support in the back part of the horse's foot, in the bars, quarters, and heel, tends to be a bit more substantial than in the sole; perhaps this means the toe area is less resistant to movement of the coffin bone.

FIGURE 3

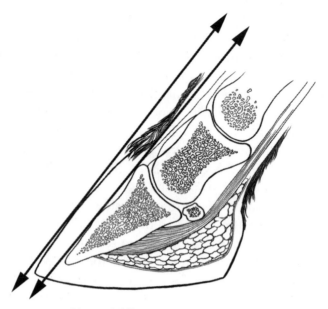

Normal Alignment

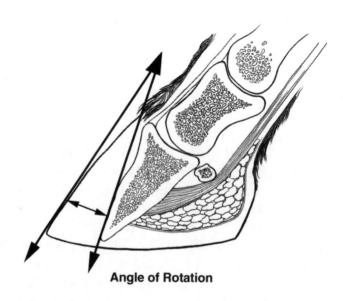

Angle of Rotation

*Diagrams based on X rays showing a normal foot
and one with a rotated coffin bone.*

In other, more severely affected horses, the entire coffin bone separates from the hoof wall and "sinks" toward the ground. In these horses, it's generally presumed that the laminar attachments have let go pretty much all the way around the hoof. These horses can be particularly difficult to treat because they essentially lose all support inside the hoof.

The second mechanical problem associated with laminitis is instability that may develop within the structures of the horse's foot itself. More simply stated, in a horse with an unstable foot, the coffin bone might be able to move around inside the hoof when the horse bears weight on its leg. Of course, these problems don't affect only the coffin bone. Rather, all of the structures of the foot become involved. Such instability may also be associated with the pain caused by laminitis.

Problems with Foot Growth and Metabolism

Problems with foot growth and metabolism can run the gamut from very mild changes, to a hoof that comes off of the foot. In between these two extremes are tremendous variations in individual horses, most likely based on the severity and distribution of the changes to the hoof structure, the length of time that the problem has been ongoing, and even the various things that have been done to the hoof in an effort to treat it. Changes in the hoof that occur as a result of laminitis may be as mild as a widening of the *white line* that's seen on the bottom of the hoof, to a flattened sole, to a slightly upright foot. Rings may become prominent in the hoof wall, starting at the coronary band and growing down (no one's really sure what causes them). When viewed from the side, the hoof wall at the toe may start to look hollowed or rounded inward, most likely due to metabolic abnormalities in the cells from which the hoof grows. More severe abnormalities of the cells of the hoof wall can lead to other problems. Indeed, sometimes there is so much cell death that effective healing of the hoof tissues simply can't occur.

Problems with Blood Vessels

In horses with chronic laminitis, damage to blood vessels may contribute to any number of problems. Most simply, an inadequate blood supply means that the tissues of the foot can't heal. In it's worst form, it means that tissues die. However, in some affected horses, new blood vessels may try to grow in an effort to supply blood to the healing foot. Unfortunately, these new tissues may actually make the foot unstable, since they are soft and flexible and don't give any firm support. Furthermore, a loss of blood vessels also means a loss of the normal cushion that those blood vessels provide. This may be one reason why some horses with laminitis never get over their pain.

Infection

Finally, the foot of the horse affected with laminitis is more prone to infection and the resulting problems. Normally, the horse's foot lives in a sea of microorganisms, but the healthy hoof is a virtually impregnable capsule and resists infection. However, in the damaged foot, the hoof is more prone to tearing and cracking and, as a result, may be more likely to be infected. Infection can be seen in any area of the foot, and show up as such things as separation of the hoof wall (known as *"seedy toe"* or *"white line disease"*), chronic thrush infections or abscesses.

Now that you've got some idea of what's happening inside of the foot of the horse with laminitis, the next thing that you'll need to know is how to recognize it. That's the subject of the next chapter.

Dr. Chris Pollitt is the Reader in Equine Medicine and Director of the Australian Equine Laminitis Research Unit at the School of Veterinary Science, The University of Queensland. He was a general practitioner for 10 years until being awarded a PhD degree by The University of Queensland in 1983 for his research into the blood types of horses. Dr. Pollitt has presented his research on the structure, function and diseases of the horse's foot (in particular, laminitis) at conferences worldwide since 1990. He is the author of a color atlas, The Horse's Foot. *This book contains text and over 400 color photographs of Dr. Pollitt's work and was published by Mosby-Wolfe, UK, in December, 1995. Dr. Pollitt is an enthusiastic amateur farrier, and actively competes in endurance riding on horses shod by himself. In 1991, he was the Australian Heavyweight Points Rider and Heavyweight Distance Rider of the Year. Dr Pollitt was inducted into the American Farriers International Hall of Fame for achievements and contributions in advancing equine foot care.*

Diagnostics

In general, the sooner that you take care of a problem, the less trouble it will ultimately be. This is pretty much the case in laminitis, although some cases do not do well no matter how quickly you recognize them and others probably get better *in spite* of treatment. Still, in order to give your horse a fighting chance for a full recovery if it develops *acute* laminitis, you should learn to recognize the common signs that may indicate that your horse is in trouble. Similarly, you should become familiar with the signs of *chronic* laminitis, so as to be able to try to work with your veterinarian and farrier in an effort to get your horse back to as close to normal as possible.

History

Veterinarians often get clues about what's happening in a horse that they're currently attending by getting information about what's happened to it in the past. Clinical histories can give valuable insight into what a horse's condition actually is. In addition, the history of every horse with a suspected case of laminitis needs to be examined to try to come up with the underlying cause. Even though you won't ever know the underlying cause of laminitis in a few horses, most of them occur because of the existence of some other problem, and that problem has wound up making your horse's feet sore.

So to what questions do you need to know the answers? Here are a few to consider.

1. Has your horse been out in grass pasture? Free access to lush, growing pasture is a common cause of laminitis.

2. Is your horse overweight? Obesity is frequently associated with the problem.

3. Has your horse been seriously ill recently? Laminitis may follow any number of medical or surgical conditions.

4. Have there been any recent changes in your horse's living circumstances? Have the stall shavings been recently changed to a different brand or substance? Has your horse had the opportunity to break into the feed room?

5. Has your horse ever had laminitis before? Horses that have had laminitis in the past are prone to get it again.

6. Has your horse received any medication? Some medications are thought to be associated with the development of the condition.

7. Is your horse very hairy? Horses, in particular older horses that develop a long, curly hair coat, the classic sign of a pituitary growth (also known as equine *Cushing's Syndrome*) are more likely to develop laminitis than are other horses.

8. Has your horse been recently shod? Trimming the feet too short or improperly placed horseshoe nails can cause inflammation of the foot.

It will really help your veterinarian if you can be as thorough and complete in answering any questions that she or he might have. In addition, try to think of things that may have recently occurred that are out-of-the-ordinary for your horse. Even if you're not sure if something is significant, little clues may be quite helpful in trying to figure out what's going on.

CLINICAL SIGNS OF ACUTE LAMINITIS

People routinely ask veterinarians if they wish that their patients could talk. Actually, in most cases, it's great that they can't talk. If they could, you'd probably have to listen to them complain about their feed, their exercise schedule, their new saddle that doesn't fit, the weight that you've recently gained, and more. Still, horses often do manage to communicate that something is wrong with them, you just have to know how to listen.

With laminitis, there are some characteristic clinical signs that will let you know that something is wrong. You don't necessarily have to know *what* is wrong; you need to know that *something* is wrong. You will recognize that there have been changes in your horse's normal appearance and/or behavior. And, if you see any of the characteristic clinical signs of laminitis, descriptions of which follow, you should call your veterinarian immediately.

Stance

Usually, the most immediate tip-off to the fact that your horse may have a problem is that it may change the way that it stands and/or moves. The severity of these changes may vary tremendously, but the classic description of a horse with laminitis is one that prefers to stand in one place, shifting its weight from one side to the other, with its front feet a bit out in front and the back feet underneath the body (Figure 4). It stands like this so as to try to get some weight off of the front feet, which usually hurt a great deal. However, this classic picture isn't the same for every horse with laminitis. A horse might not fit the classic picture if only one foot is affected, such as happens when a horse tries to support all of its weight on one foot only, say, after a serious injury to the opposite leg. Although it's unusual, and occurs generally only in the most serious cases, the back feet may also become involved along with the front (the back feet are never solely affected). Horses with laminitis in all feet tend to stand with the fore and hind feet bunched together under their bellies, sort of like a circus elephant on a stand. Still, in general, if a horse has laminitis in both front feet, as is most common, you'll see some variation of the classic stance. And the worse the problem, the more obvious are the changes in the stance.

The stance of the horse with laminitis can also change as the severity of the problem changes. For example, a horse with mild laminitis may look pretty normal if it's standing on soft ground and it doesn't have to move around a lot. But, as the disease gets worse, or as the ground gets harder, changes in the stance may become more obvious. On the other hand, in some severely affected horses, their stance may seem pretty normal, possibly because there's been so much damage to the feet that the horse can't feel them. In the most severe case, the horse may not even want to get up and stand at all.

The bottom line is the horse is going to stand in the way that

FIGURE 4

*The front feet are forward and the back feet are underneath,
thus assuming the classic stance of a horse with laminitis.*

makes it the most comfortable. Still, while a horse that adopts the classic, foot-forward stance should be looked at closely to see if it has laminitis, the absence of such a stance doesn't mean that the horse is free of the problem. Importantly, since the best chance for successful treatment occurs when you get to the horse early, you shouldn't wait for dramatic stance changes to come about before you express your concerns to your veterinarian that something might be wrong.

Gait

By the time the horse's stance is affected, its gait is usually pretty messed up, as well. Horses with laminitis tend to try to move with their feet out in front of their body. They move this way because they're trying to put most of their weight on the heels of the foot. Some affected horses look as if they are walking on eggs. They also seem to hate turning. This is most likely because turning causes more weight to be put on the foot that's being turned upon, and also causes the structures inside the foot to twist. Gait changes can vary from horse to horse, from a horse that's just a little bit sore when asked to turn, to one that's reluctant to put any weight on the problem feet at all.

Although the lameness gait patterns in horses with laminitis can vary considerably, you'll usually find that both front feet are involved. But don't forget, even if a horse shows its lameness only in the front feet, you still want to check the ones in the rear. As noted earlier, the back feet may become involved in laminitis but the signs usually aren't as dramatic as front-foot laminitis, particularly in less severe cases. This is probably because the horse bears much more weight on its front feet than its back feet; the extra weight makes the front feet extra sore. But horses with laminitis in the back feet tend to move strangely. They often tend to rock forward on their front feet and walk really quickly. They also may tend to pick the back legs up rapidly and in an exaggerated fashion, as if they're walking on hot ground. They probably move in this way to try to keep from rolling over their poor sore toes.

The gait of the horse with laminitis may also change depending on the surface that the horse is asked to walk on. As you might guess, sore-footed horses really hate walking on hard or uneven surfaces. They're generally much happier in a deeply bedded stall, on soft ground, or in deep sand.

Carefully watching and recording how your horse with laminitis moves is a pretty good way to assess how it is doing, as well. There's usually a reasonably direct relationship between gait improvement and improvement in the disease condition. Otherwise stated, the lameness associated with laminitis tends to get worse as the disease gets worse. On the other hand, you'll have some idea that your horse is getting better when it starts moving better. But don't put too much stock into how your horse looks on one particular day, as some horses may have their ups and downs before finally recovering.

Digital Pulse

The two arteries that run down the back of the pastern can give you a clue as to whether or not there's inflammation in your horse's foot. Learning to feel the pulse is useful for the diagnosis of many lameness conditions in the horse, including laminitis. Take some time to find the normal pulse and get used to taking it. You'll be more likely to feel when it's abnormal, should an abnormality occur. If you've ever hit your thumb with a hammer you know that it throbs and hurts; you can feel that same throbbing in the character of the pulse of a horse with laminitis. If you suspect that your horse has laminitis, it's important to check the pulse in all four of the horse's feet.

However, just because the pulse to your horse's foot may be a bit stronger than normal, don't panic. There are plenty of other things that can cause the character of a horse's pulse to increase besides laminitis. For example, the pulse can feel stronger after a horse has been worked on hard ground, when it's excited, or if it has an abscess in the foot. It's probably also a good idea to check the pulse in an individual limb several times, to make sure of your finding. If you do think that the pulse to your horse's foot is increased above normal, and particularly if you have some additional clinical signs or other abnormalities, don't be afraid to call your veterinarian and ask about it.

Hoof Temperature

When any area becomes inflamed, it also gets warmer. This is something that can actually be measured. That being said, trying to pick up heat in a hoof can be pretty subjective, and it's really hard to do if the weather is also warm. So, if you're concerned that your horse's feet are warmer than normal, and particularly if there's some lameness or gait abnormality at the same time, it's probably worth calling your veterinarian to see if he or she feels that it's necessary to come out and conduct a thorough examination.

Pain on Hoof Examination

While the way that the horse moves or stands may let you know that he's in pain, it's important to look at each foot to try to get a feel for how much pain is occurring in each foot and to try to determine the area of that foot from which the pain originates. To make such determinations, your veterinarian may try to manipulate the foot to see if there is soreness in response to those manipulations. He or she will also undoubtedly use a device called a hoof tester. A hoof tester is simply a big set of pincers, which squeeze the hoof. In response to hoof tester pressure, horses with laminitis often show pain over the sole, along the rim of the underlying coffin bone, or right in front of the tip of the frog. However, pain may not be confined to the sole and some horses with laminitis may have pain to hoof-tester examination in other areas, as well.

Hoof-tester examination is certainly not foolproof. Some horses with severe laminitis may have a foot that doesn't react to them at all. This may occur in horses in which there has been so much damage to the foot that it's become essentially numb. It's also not possible to do a hoof-tester exam in every horse. Some poor horses may be so foot sore that they refuse to stand on one limb while you try to hold up the other. About the only thing that you'll do in trying to get the foot of this kind of horse off the ground is hurt your own back!

SIGNS OF CHRONIC LAMINITIS

Horses that have had laminitis for some time may have any or all of the signs that are associated with *acute* laminitis. However, in the horse with *chronic* laminitis, the normal anatomy of the hoof has been disintegrating, even though you most likely haven't been able to see it. Still, over time, the hoof of the horse with chronic laminitis will most likely begin to show the results of that disintegration.

Even if you can get the horse with laminitis to recover, changes that may occur in the hoof as a result of the condition may mean that you're going to become very well acquainted with your veterinarian and farrier, who are going to attempt to eventually set things straight.

Changes in the Coronary Band

When the laminae that hold the coffin bone inside the hoof let go, the bone drops down into the hoof. This may result in a noticeable depression that you can feel at the top edge of the hoof wall. At first, you may only feel this right over the front of the hoof. However, the depression can become quite a bit more extensive, extending around to the sides of the hoof, or even the heels. In such cases, the prognosis for the horse is usually pretty poor, because it means that the laminar attachments have let go nearly all the way around the hoof. In severe cases, the skin may even start to separate and the foot may start to leak sticky, yellow blood serum. This is usually a really bad sign.

Changes in the Sole

It may take no more than a few days for changes in the sole of the hoof to occur. The most common change is a loss of the normal concave nature of the sole, which may occur as the structures above start dropping down. In less severe cases, the sole may start to flatten. In more severe cases, the sole may even develop a convex shape, with a bulge pushing down toward the ground. The bulge may be accompanied by changes as minute as little cracks over the bulge or, in the most severe cases, the coffin bone may follow the bulge right out the bottom of the hoof.

The sole may also develop a ring of red or purple bruising around it. This generally means that there has been a good deal of internal damage and bleeding from the sensitive tissues of the sole. Sometimes, you run across areas where there is old blood; this is just another indication that there have been some bad things going on inside your horse's hoof.

Changes in the Hoof Wall

If, after an episode of *acute* laminitis, the bone of the foot has changed its position, the horse may end up having a deformed hoof. At the front of the hoof, where the most damage usually occurs, growth of new hoof may end up being slower than normal. At the same time, at the back of the hoof, where minimal damage may have occurred, hoof growth may proceed at its normal pace. When viewed from the side, this can give the wall a cupped, or "slippered" appearance, and the heels can get quite long.

As laminitis progresses, you might also notice prominent rings developing in the normally smooth wall of the horse's hoof. If you look closely, you'll see that the normal hoof has growth rings around it that are parallel to the coronary band. But in the horse with *chronic* laminitis, the growth rings aren't parallel to the coronary band anymore and they tend to get closer to each other at the toe, where the hoof growth is slowest (Figure 5). In addition, if the bone drops inside the hoof, it moves the growing hoof tissue as well, resulting in a kinked or grooved appearance to the hoof wall.

In *chronic* cases, the hoof wall itself may start to separate. This can lead to additional pain and infection; it may become yet another problem for your veterinarian and farrier to deal with.

X Rays (Radiographs)
It's usually not all that hard to come up with a diagnosis of laminitis based on the history, physical examination, and the clinical signs of disease. But the information that you get from X rays usually helps you make the final decision about how bad things are, how long the problem has been going on, what sort of treatments you might consider undertaking, and the prognosis for the future. Taking X rays of the feet of a horse with laminitis—especially a side (lateral) shot showing the relationship of the bone to the hoof—is an absolutely critical part of the whole process.

There are a couple of times that you might consider taking X rays. One would be as soon as the clinical signs of the disease show up. In this way, you've got something that shows you where the bones were at the start of the mess that your horse is in. It's also important to x-ray the horse as things progress so that you can see if you're helping the horse with the treatment that's being provided. If all is going well, you'll like knowing about it. Conversely, if it's not, the information about what's going on inside your horse's foot will help you make important decisions. Over time, you'll end up with a file in your veterinarian's office that will show the progress that your horse has made, good or bad.

FIGURE 5

*Over time, abnormal rings, a long toe, and a high heel
may develop in the horse with chronic laminitis.*

Early on in the course of laminitis, X rays may not show your veterinarian all that much. Ideally, they'll *never* show any changes from normal, but you do need to know. The earliest changes seen on X rays are an increase in the distance between the hoof wall and the coffin bone. This isn't necessarily easy to see. It's the sort of thing that can be measured if the changes are obvious, but it can be difficult to appreciate subtle changes. Your veterinarian may look for other things on the initial X rays, such as the appearance of dark lines that indicate that there has been hoof-wall separation, or changes in the position of the coffin bone, which indicate movement of the bone inside of the hoof.

With time, the X ray changes of laminitis may become more dramatic. As you know, in *chronic* laminitis, the top of the coffin bone may start to rotate away from the hoof wall, pulled by the deep digital flexor tendon. As this happens, the sensitive tissue of the horse's sole may get squashed and the tip of the toe may appear closer to the ground. This pressure may cause that tissue, as well as the tip of the bone, to die, and such changes can be seen on the X rays, as well.

In any case, the degree of rotation, as noted on X rays, appears to be of some importance, at least from the standpoint of predicting how things are going to go. It's generally felt that horses with less than 5.5 degrees of rotation of the coffin bone have a pretty good chance of getting back to full athletic function. On the other hand, horses with more than 11.5 degrees of rotation have a rather poor prognosis. These horses generally can't go back to doing their previous jobs, may be crippled, or may have to be put to sleep because there is no way to control their pain. These aren't hard and fast rules—some horses can come back from dramatic rotations—but they are useful guidelines in helping you make your decision as to which treatment to use and for how long to pursue it.

In some unfortunate horses, there is so much damage that the bone doesn't really rotate. Instead, virtually all of the hoof attachment lets go. In these horses, the bone sinks in the hoof, toward the ground, (and the horses are generally referred to as "sinkers"). Sinking can also be seen on X rays. These horses are usually in a lot of pain and the outlook for them is usually not good.

But your veterinarian is looking for more than changes in the position of the coffin bone when he or she is looking at X rays of your horse's feet. Changes in the shape of the coffin bone, in which the bone can take on the appearance of the tip of a ski, or fractures from the edges of the bone, or signs of inflammation and/or infection may take weeks to show up. When such changes are present, they generally indicate that things are well along in the disease process. In the most severe cases, dark gas lines may appear on the X rays. These lines, which look like black streaks in the otherwise gray wall of the hoof, indicate that the hoof and the bone are separating or that there may be infected tissue in the hoof. If the disease process has been going on for a long time, the coffin bone can even start to demineralize, which can cause some very weird-looking X rays, as well as a very lame horse.

The bottom line is that X rays are essential to diagnose and monitor the progress of a horse with laminitis. They're also very important in directing the effort of taking care of the horse's foot (more on that in Chapter Six). Don't think of doing without them.

Causes

With Philip Johnson, BVSc (honors), MS,

Diplomate ACVIM, MRCVS

Another irritating aspect of laminitis is that it can occur in association with an almost unbelievable number of factors. Indeed, laminitis appears to represent a common, end-stage event that can have many different causes. As you know, the details of why and how different circumstances lead to what is easily and readily recognized as laminitis are the subject of intense research and active debate. Still, it's important to know about these various causes so that you can try to get to the bottom of your horse's problem and, if possible, get rid of it. It's also important to know about these various causes from a prevention aspect. (See Chapter Eight for more on preventing laminitis).

Still, even though laminitis is often associated with a whole lot of different conditions and causative factors, there have been very few useful published surveys that have been able to show a true cause-and-effect relationship. Much (too much) is based on speculation, opinion, or stories, rather than on hard facts.

To make things even more confusing, when laminitis happens following some other event, you can't necessarily even conclude that the event caused the laminitis. There might be an association with a lot of unrecognized or not so obvious "other" factors. For example, some horses may have pre-existing long-term, but currently inactive laminitis. (Here's something to remember—once a horse has laminitis, it always has laminitis, even if it's not obvious.) These horses may be perfectly fine, and you might be completely unaware that there was ever a problem. However, they may be prone to relapses and these can be confused with primary laminitis from another cause.

Still, the various causative circumstances that are associated with the development of laminitis can be divided into several subsets, which are discussed in this chapter. The most common and most widely recognized of these occur in horses that develop laminitis as a result of some sort of *disturbance to the gastrointestinal tract*. In a few published studies, gastrointestinal disease is associated with approximately 50 percent of the "new" cases of *acute* laminitis. The gastrointestinal disturbance doesn't have to be so great as to result in obvious clinical signs like colic or diarrhea; in fact, you may not even know that such a disturbance has occurred!

The other common circumstance associated with laminitis is the release of as yet unidentified bacterial toxins that normally inhabit the horse's large intestines. It's likely that many causes of what appear to be so-called "new" cases of *acute* laminitis with no obvious cause can be attributed to either disturbances in the gastrointestinal tract, or stress, with its associated changes in glucocorticoid hormones, or the presences of pre-existing, but unrecognized disease, or any combination of the three.

Still, even with all the confusion, it is possible to try to categorize the various causes of laminitis. Keep in mind that there aren't any hard lines when it comes to these categories and sometimes more than one causative factor might play a role.

DIGESTIVE UPSETS

Certainly, the most well known cause of laminitis relates to a horse eating excessive amounts of easily digestible carbohydrates. Carbohydrates are the sugars, starches, and fiber that make up a large part of the plants and grains that horses eat. Such things as hay or coarse pasture may contain carbohydrates, but such fibrous materials are not easy to digest, so they're not typically associated with the development of laminitis. Ingestion of too much easily digestible carbohydrate—which contains lots of sugar and starch, but not so much plant fiber—may occur when a horse eats too much grain, such as might happen in a stabled horse, and particularly when it's not used to getting much of it. Lots of lush grass can also be a culprit. Even though grass is "natural" and nutritious, in natural conditions, horses would normally be moving around and exercising, not grazing and gorging themselves for hours in one fenced-in spot. Even simple diet changes, especially from all-forage diets to ones that include higher amounts of grain, can lead to laminitis. Simply stated, from whatever the source, large amounts of easily digestible carbohydrates are clearly not a good thing for the horse.

The "classic" laminitis story is about a horse that breaks into a grain room or feed bin and is found munching away to its heart's content. Under such circumstances, the horse's intestinal tract can get overloaded with a large amount of easily digested carbohydrates. In fact, carbohydrate overload is such a good way to make a horse get laminitis that it's the primary way that researchers who are studying the condition get horses to develop it.

Excessive consumption of fresh grass can also cause laminitis. Under certain conditions (especially springtime, with its cold nights and sunny days) grasses manufacture high concentrations of easy-to-digest carbohydrates called fructans. In addition, these lush grasses are low in fiber. The combination of high carbohydrates and low fiber seen in lush grass is exactly the same formula that occurs with grains, and excessive consumption of grass causes the same problems. In fact, overeating lush grass was by far the single most common cause of laminitis according to a United States government survey published in 2000.

The horse's large intestines have very important digestive functions. Digestion is the process by which food is broken down into smaller molecules that can be absorbed through the intestinal walls into the bloodstream. Those molecules are then delivered to the tissues of the horse's body by its circulatory system. To perform its functions successfully, the large horse's intestines rely on a delicate balance of bacteria and an intact intestinal lining. When lots of grain or other easily digested carbohydrate gets to the large intestines, the intestinal environment becomes more acidic. An acidic environment may cause many of the normal bacterial residents to die and release bacterial toxins, which then may be absorbed through the intestines and into the circulation. Newer theories suggest that when the intestinal environment changes, bacteria that produce laminitis-triggering factors may grow at the expense of others. An acidic environment may also damage the lining of the intestines, which normally protects the body from the intestinal contents. Damage to the intestinal lining may allow for easier passage of bacterial toxins and certain as yet unidentified factors through the gut and into the circulation, where they may cause harm when delivered to certain body tissues, and most especially those of the feet.

The horse's intestines are very sensitive. When they are adversely affected—from just about any cause—laminitis can result. For example, laminitis can follow a bout of severe colic. The word *colic* simply means pain coming from the horse's abdomen, and there are dozens of causes. Fortunately most cases resolve without much ado, but those horses with more severe colics—ones that require surgery, or that are associated with inflammation of the horse's gastrointestinal tract—are prone to develop laminitis. And, the more severe, or the greater the complications, the more likely it is that laminitis will occur.

Numerous other conditions can cause inflammation, and/or infection of the gastrointestinal tract. Two examples are: infectious diarrhea caused by the salmonella bacteria or the inflammation of the colon (colitis) that is caused by such diseases as Potomac Horse Fever. These diseases can severely damage the gut and impair the normal protective barrier that the gut provides, at the same time as the contents of the gut are disturbed. In addition, with such diseases, pain, stress, and circulating bacterial toxins further compound the situation. As such, you've got an ideal vat in which you can brew a case of laminitis (discussions of pain and stress, and circulating toxins, follow).

Even seemingly innocuous things have the potential for upsetting the horse's gastrointestinal tract, although it's not always clear why. For example, simply changing a horse's diet is associated with an increased incidence of colic, and colic is associated with laminitis. Even deworming has very occasionally been associated with laminitis in the horse though it's important to recognize that this extremely small risk is no reason to avoid this important part of good horse management.

VASCULAR CAUSES

Any factor that interferes with the circulation of blood to the laminae of the feet has the potential to cause laminitis. This can be seen in any number of conditions, particularly those in which the circulating blood volume isn't sufficient to meet the horse's needs. For example, the circulation becomes impaired when a horse goes into shock (shock being a serious derangement of the horse's normal metabolism). In shock, there is generally either a loss, or an inappropriate distribution, of blood. Laminitis may result. Similarly, with severe disease of the gastrointestinal tract, massive amounts of fluid can be lost in the form of diarrhea, and this has the potential for adversely affecting the circulation to the horse's feet because there's only so much fluid to go around. Even such events as failure to maintain normal blood pressure when a horse is subjected to general anesthesia can cause a horse to develop laminitis.

Other diseases affect the circulation and blood vessels more directly. For example, purpura hemorrhagica is an inflammation of blood vessels that's caused by rampant inflammation. It's an unusual, though serious complication that's seen especially after cases of infection with *S. equi* ("strangles"), but can also be seen following just about any disease in which the immune system is over stimulated. Other conditions can cause problems such as abnormalities of blood clotting or destruction of red blood cells. All can result in laminitis.

BACTERIAL TOXINS

Bacterial toxins normally found in the cell walls of certain bacteria may also play a role in the development of laminitis in some horses. At one time, some of these toxins were thought likely to be a primary cause of laminitis, but this is not so widely believed now. Current thought is that circulating bacterial toxins may lead to biochemical and cellular events that end up compromising circulation and destroying local tissues, which is exactly what you don't want to have happen in a horse's foot.

Bacterial toxins may enter the horse's bloodstream from any number of sites, as a result of any number of disease conditions. For example, if there's damage to the gut, toxins may be more likely to get into the circulation and do harm. Pneumonia and pleuropneumonia (also known as "pleuritis" or "shipping fever") are infections of the lungs and/or chest cavity, respectively, that may develop after horses are transported. If the diseases can't be brought under control, they may result in laminitis. Infections in large tubular organs such as the uterus, which may result if the placental membranes are not expelled within a few hours after birth, can result in laminitis. Presumably, the infected areas in the preceding diseases serve as a storehouse for bacteria, bacterial toxins, and as-yet-unidentified laminitis trigger factors.

OTHER TOXINS
Laminitis can be triggered by any number of items that the horse eats or may be exposed to. In such cases, laminitis is a reaction to a poison. Some of the most commonly recognized toxins include:

Black Walnut Shavings
The heartwood of the black walnut tree contains a toxin that, even in tiny amounts, can cause horses to develop laminitis. This usually becomes a problem when the shavings used to bed horse stalls contain black walnut. Horses have developed clinical signs of toxicity when as little as 5 percent of the wood shavings are black walnut. The freshness of the shavings might also be a factor insofar as their ability to cause problems in a particular horse.

The toxin in black walnut shavings is presumed to be absorbed through a horse's skin, although it's certainly possible that a horse could get a big oral dose if it ate the bedding. In horses that do get a significant dose, laminitis may occur in as little as 12 to 24 hours after exposure. Other signs of a problem with black walnut shavings include limb swelling, depression, colic, fever, and loss of appetite, which may precede the development of laminitis.

Molds and Fungal Toxins
Molds, and the toxic by-products that they can produce, have been associated with the development of laminitis in horses. The most common source of these is found in feed. Obviously, you should always feed your horse the best quality feed available so as to avoid problems associated with poor-quality feed.

Endophyte-Infested Fescue?
While it's not completely clear, there may be a link between laminitis and tall fescue grass that is infected with a type of fungus called an *endophyte*. Endophyte is a fungus that lives inside a grass plant. Fescue is a common pasture grass, particularly in the eastern United States. It's certainly known that endophyte-infested fescue causes

all sorts of problems in late-term mares, such as delayed foaling and lack of milk production. But there's some evidence that it may also increase the risk of laminitis by virtue of increasing the tendency of blood vessels to spasm when the horse is stressed, excited, and/or diseased, leading to decreased blood flow to the foot.

Kidney Failure

The horse's kidneys have very important functions in removing many of the body's accumulated by-products of the normal metabolic processes. Although it's not common, if the horse is unable to remove normal body wastes due to *impaired kidney function,* it may also be at increased risk for laminitis.

Rhabdomyolysis ("Tying Up")

Severe muscle cramping may occur in a horse after the onset of heavy exercise, particularly after a period of rest. This is commonly referred to as "tying up." Like laminitis itself, the condition is also associated with feeding large amounts of grain. While the cramping associated with tying-up is painful, the condition can cause muscle cell damage in severely affected horses. In some cases, that damage is extensive. Affected muscles cells release their muscle pigments, which can then harm the horse's kidneys. The resulting direct injury and the accompanying stresses, fluid abnormalities, and pain may cause some horses to develop laminitis.

Drug Reactions

Laminitis has been reported as an unfortunate side effect after the administration of some drugs, for example, quinidine, which is used to control abnormal heart rhythms. Fortunately, such side effects appear to be rather rare. Unfortunately, they are also totally unpredictable. However, one type of drug, the *corticosteroid anti-inflammatory agents,* as well as related compounds that occur naturally in the horse, is of particular concern and is worthy of more detailed scrutiny (see below).

Corticosteriod-Associated Laminitis

Corticosteroids are a group of potent anti-inflammatory drugs that are tremendously useful for treating a large number of medical conditions. Unfortunately, giving them to horses also seems to be associated with an increased risk of developing laminitis. It's not a direct relationship; researchers haven't been able to cause laminitis simply by giving horses corticosteroids. But they may increase the risk of laminitis when other stress factors, such as showing, heavy training, shipping, inactivity, illness, or obesity, occur. It's not at all known why corticosteroids may increase the incidence of laminitis, although it has been shown that they tend to increase constriction of the blood vessels. As you know, constricted blood vessels may reduce blood flow to the feet.

Certain corticosteroids appear to be more likely to cause problems than others. In particular, *triamcinalone* and *dexamethasone*—two long-lasting steroids—appear to increase the risk for laminitis, possibly by causing spasms in the little muscles of the blood vessels in the horse's foot. The likelihood that corticosteroids will cause laminitis also appears to be related to the dose of the drug given, and for how long the horse is treated.

That's not to say that corticosteroids are killer drugs, however. The vast majority of horses that receive corticosteroids never have problems with them and the extremely small risk of a horse developing an unforeseen reaction is no reason to avoid the use of these drugs altogether. Of course, like all drugs, corticosteroids should be used judiciously and appropriately and on the advice of your veterinarian.

Cushing's Syndrome

In addition to corticosteroids from external sources, two laminitis-producing conditions appear to be related to high levels of the horse's body's own corticosteroid hormones, known as *glucocorticoids*. The first condition is a benign tumor of the horse's pituitary gland, commonly called equine *Cushing's Syndrome*. Pituitary tumors are most commonly seen in older horses. The most consistent sign of equine Cushing's Syndrome is a long, curly hair coat that resists shedding. Other signs include weight loss (in spite of a good appetite), poor condition, increased drinking and urinating, delayed wound healing and, of course, laminitis. The laminitis associated with Cushing's Syndrome is most likely as a result of higher-than-normal circulating levels of glucocorticoids.

Obesity-Associated Laminitis

The second condition that may be related to excessive levels of the horse's body's own steroids is also seen in mature horses. In these horses, veterinarians are starting to recognize a peculiar syndrome of *obesity* and laminitis. These portly beasts have a characteristic distribution of body fat that typically involves the neck and the rump. Affected horses commonly accumulate extra fat in the crest of the neck, at the top of the rump, and in the sheath area of the male. It's often quite difficult to get these horses to lose weight by simply reducing their feed; they're commonly known as "easy keepers." Broodmares affected with this problem typically show abnormal heat cycles and they are notoriously difficult to breed successfully.

For some time, many people have wrongly said that this fat-horse syndrome occurs because of problems with thyroid metabolism (hypothyroidism). That's probably because lowered, circulating thyroid-hormone levels can often be demonstrated when blood tests are performed on horses that have obesity-associated laminitis. However, there's really not anything wrong with the thyroid gland in these animals. It's just that the increased production of the body's glucocorticoids can interfere with a hormone produced in the horse's pituitary gland at the base of the brain that is responsible for regulating the thyroid gland. Thus, low-thyroid hormone levels in these chubby chargers are most likely a secondary—and maybe clinically irrelevant—effect.

Nor do these fat horses with sore feet usually have any evidence of equine Cushing's Syndrome. Tests for Cushing's Syndrome are most often normal, as is the pituitary gland of affected horses when they are looked at on a post-mortem examination.

What these horses do show are some signs that suggest that increased levels of glucocorticoids have affected them. Researchers at the University of Missouri have speculated that increased glucocorticoid production by tissues that don't normally produce the hormone might be responsible for the clinical appearance of obesity-associated laminitis in horses. Their preliminary studies suggest that this syndrome in horses is very much like a condition called "central obesity" in people. Horses with obesity-associated laminitis, and people with central obesity, both tend to have increased blood pressure (hypertension) and insulin resistance (elevated insulin levels). Other common shared abnormalities include reduced fertility, trouble losing weight, and an abnormal distribution of body fat stores. One thing that may play a role in development of the condition is nutrition, especially early in life. Accordingly, it's important not to overfeed your horse from the get go. Further investigation is ongoing to determine whether other similarities exist between the human and the equine conditions.

Other Hormones and Stress

Glucocorticoids are not the only hormones that are associated with the development of laminitis. Abnormal reproductive hormones, in particular, *abnormal levels of estrogen* in the blood, such as may occur with ovarian disease, may be associated with laminitis. Fortunately, this sort of thing appears to be rather rare. Even *stress,* the type that can accompany disease, or that can occur as a result of long-distance travel, for example, causes higher levels of certain hormones associated with spasticity of blood vessels.

GLUCOSE PROBLEMS

Some fairly recent work suggests that abnormalities in the horse's body's use of glucose—the main sugar that's found in blood and the main source of energy—is at the root of some cases of laminitis. The cells that make hoof have a strong appetite for glucose. If glucose isn't available, or circumstances mean that the cells aren't getting enough glucose, laminitis could be the end result. There may also be some connection with glucose and glucocorticoids; these hormones tend to reduce the ability of the cells to take up the needed sugar.

This appears to be a particular problem for ponies. In fact, ponies with laminitis tend to be somewhat intolerant of glucose when compared to horses. It's interesting to speculate whether this is the reason that ponies seem to have an increased risk for laminitis.

LAMINITIS AND INFLAMMATION

Laminitis causes inflammation in the horse's foot, so it stands to reason that just about anything that can cause inflammation in the foot might potentially result in laminitis. Thus, aggressive cutting and trimming of the hoof, or problems with the hoof itself, such as untreated abscesses or deep hoof infections, at least have the potential for causing more complicated problems. In the same manner, any sort of trauma to the hoof can serve as a trigger for laminitis. For example, horses that are forced to do heavy work on hard ground, or spend excessive amounts of time on hard floors may be at particular risk.

MECHANICAL CAUSES

As it relates to laminitis, the word *mechanical* means that the stresses that are applied to the horse's leg exceed the ability of the horse's leg to withstand them. Mechanical insults are like the proverbial straw that broke the camel's back. The system can stand up to a good deal of stress but when it's overloaded, something has to give. Sometimes, that something is the supporting tissue of the horse's hoof.

Mechanical causes of laminitis are sometimes referred to as "support laminitis." As such, it's most commonly seen when one hoof is bearing all the weight of the horse. This usually happens when severe primary lameness pain or a painful surgery somehow compromises one leg of the horse and causes it to bear excessive weight on the opposite leg. Unfortunately, under such circumstances, the system may not be able to hold up under the strain.

Other mechanical forces may come into play, as well. For example, laminitis is very rare in light, little Miniature horses, but all too commonly severe in big, heavy draft breeds. Similarly, laminitis is almost unheard of in light foals and yearlings, rather, it's a disease of larger, adult animals. A very small hoof on a very large horse may take more pressure than a same-sized horse with a larger hoof; this could also lead to a predisposition to laminitis.

DISCREDITED "CAUSES" OF LAMINITIS

As important as it is to know something about what *does* cause laminitis, it's probably also a good thing to know something about what *doesn't* seem to cause laminitis (even though some people may believe that it does). In fact, some of the "traditional" causes of laminitis have been somewhat discredited in recent years.

Hypothyroidism

Low levels of thyroid hormone (hypothyroidism) are commonly thought of as being a trigger factor for laminitis. The association has been made because fat horses tend to get laminitis and, in other species, low levels of thyroid hormone tend to result in obesity.

However, it's becoming increasingly clear that the combination of laminitis and obesity do *not* occur simply because of a lack of sufficient thyroid-hormone production. While some of these horses may have low levels of thyroid hormone on a blood test, individual blood tests are not really significant (the levels of thyroid hormone in the horse go up and down throughout the day, and a single test only tells you where the levels are at that particular time). In fact, appropriate clinical testing, which includes a test that stimulates the thyroid gland, usually fails to support the diagnosis of hypothyroidism. Further evidence that fat horses don't have low thyroid hormone levels is provided by the fact that when you surgically remove the thyroid glands of a horse (experimentally), they get neither fat nor do they develop laminitis; rather, they get skinny and look unthrifty.

Ingestion of Cold Water

Some people assert that if a horse drinks too much cold water, particularly after exercise, it will develop laminitis. There's certainly no evidence at all that this is the case and it's pretty hard to come up with a mechanism whereby this could occur (the cold water gets warm very quickly as soon as it's inside the horse). You certainly don't want to restrict a horse's water when it is exercising—proper hydration is essential for good performance. You might consider preventing a hot horse that has finished exercising from drinking too much water too quickly so as to keep it from developing colic. But you certainly don't need to worry about the temperature of the water.

Excess Protein

"Too much" dietary protein has been wrongly implicated for all sorts of ills that beset horses. From a historical perspective, the association of grain intake with laminitis was noted long ago. People used to think that the reason that grain caused laminitis was because it contained protein, but now they recognize that it's the carbohydrate aspect of grain that's the problem.

Protein is a necessary component of the horse's diet. It's provided in virtually everything that a horse eats. Some feedstuffs, such as alfalfa hay, have more protein than is required by the horse. When given protein in excess of their needs, horses simply use it to produce the energy that's needed to run the body. Some horses that get fed excessive amounts of protein may be seen to urinate excessively (to get rid of the extra nitrogen that's comes with the extra protein) but they do not develop laminitis.

Endotoxin
Certain types of bacteria have a toxin in the walls that surround them. This toxin, known as *endotoxin,* makes normal horses very sick. Endotoxin can be absorbed by the horse as a result of a number of diseases that are related to the gastrointestinal tract. Many of these conditions can also result in laminitis. Thus, for a long time it was felt that endotoxin was likely to be a trigger factor for laminitis.

Unfortunately, at least for those that proposed the theory, infusing normal horses with endotoxin does not cause them to develop laminitis. Perhaps endotoxin is necessary to set up the events that lead to laminitis, but it is not a simple cause and effect. Insofar as you're concerned, it probably doesn't matter all that much, because the bacteria that have endotoxin in them are part of the normal bacteria contained in the horse's large intestines. But now you know.

Obviously, there are many disease conditions that can lead to laminitis. And, with so many causes, no single treatment is going to be effective in treating every case of the disease. In the next chapter, you'll read about various treatments done in an effort to lick this troublesome problem.

Dr. Philip Johnson teaches equine medicine and surgery at the University of Missouri's College of Veterinary Medicine, where he is a full professor. Dr. Johnson graduated from the University of Bristol in England in 1981, and has been specializing in the veterinary care of horses in the U.S.A. since 1983. He is board-certified in internal medicine by the American College of Veterinary Internal Medicine. He has published almost 100 articles in peer-reviewed journals, and lists equine laminitis among his primary research interests.

CHAPTER 5

Medical Treatments

With N.T. Messer, IV, DVM

The process that starts the destruction of the laminar attachments begins to operate before the first clinical signs of laminitis—that is, foot pain—become evident. By the time that foot pain is evident, damage is already occurring to the laminar tissues that keep the hoof attached to the coffin bone. Ideally, you don't want to wait until a crisis occurs to start treatment. After you've recognized that the horse has a condition that may result in laminitis, to wait and see if foot pain occurs is to miss an opportunity to prevent, or at least minimize, the damage to the foot.

As noted earlier, if you're going to have a chance of a successful outcome, it's absolutely critical that any specific problems of the horse that are associated with development of the laminitis be attended to. You can't just look at the horse's foot and forget that there's a horse attached to it. Laminitis usually develops as a result of some other process rather than as a primary disease of the foot. Thus, it's absolutely crucial that you address any of the various underlying disease processes associated with laminitis (the treatments for the various underlying conditions are too numerous to mention in this short book). If you can get on top of the primary disease, you have a much better chance of reducing damage to the foot and improving the overall chances that things are going to get better.

However, no matter when it is applied, no known medical therapy has been shown to be able to reliably stop or prevent laminitis. Any number of therapies may be used in an effort to provide some relief to the horse once it has the condition. It's fair to say that the extent of the initial problem influences how things ultimately turn out more than does the particular treatment given. Happily, not all horses that get laminitis develop all of the serious complications of the disease. In fact, some percentage of horses appear to recover with no leftover damage to the foot at all.

In general, medical therapies for laminitis need to be based on three things: 1) what's going on in the foot at the time that treatment is being provided, 2) an understanding of what the treatments that you're providing are supposed to do, and 3) any response that the horse has to the treatments that are being provided (you tend to keep doing something that seems to be helping and you probably change the course of therapy if the horse's condition is getting worse in the face of a treatment).

Thus, one strategy for laminitis treatment is directed to relieve pain and/or inflammation and to somehow affect the circulation of blood to the foot, in any number of ways. Additional therapies may be aimed at perceived or real hormonal abnormalities. Another strategy might be aimed at preventing enzyme activity or slowing down the release of the unknown factors that may trigger the problem. Of course, treating laminitis is not just a matter of prescribing medicines; treatments may include nutritional management, and, of course, foot care, to name two.

No matter what you do, keep in mind that there is really very little in the way of firm, factual data from which *any* consistent therapeutic regimen can be prescribed. Consequently, an individual horse may be treated in any number of ways and treatment *may* be "successful" (or not). It also follows that there is virtually no good evidence that any particular treatment will be effective in all situations. The bottom line is that you should be rationally—not desperately—trying to treat your horse with laminitis. With the information that follows, you should be able to do just that.

REMOVE THE CAUSE
The first consideration in the medical management of laminitis is to try to get rid of any underlying cause of disease. This probably seems obvious, but you're not going to get anywhere trying to treat the signs of laminitis if there's some disease process driving the whole thing. If your foot hurts because there's a rock in your shoe, you want to take the rock out. Taking an aspirin for the pain isn't a very smart approach. It's the same with laminitis. If there's a

definite cause you need to go after it. So, in selected cases, antibiotics might be warranted (to treat infections), or intravenous fluids might be mandatory (to treat fluid loss caused by severe diarrhea, for example). If the horse is obese, a strict diet might be in order.

On the other hand, in some horses, the cause isn't obvious, or, if it is, it isn't curable. So, for example, horses with Cushing's Syndrome can be very frustrating to treat because there's no way to get rid of the growth on the pituitary gland that's at the root of the problem. You need to keep this in mind when it comes to undertaking treatment of such horses, because if you can't get rid of the cause, you may be looking at life-long therapy, with all its expense and commitment of time.

TREAT BEFORE THE ONSET OF DISEASE

Sometimes, it may be possible to treat a horse *before* it develops full-blown laminitis. In particular, if you discover a horse after it's broken into some delicious but dangerous feed, you may have the opportunity of nipping the situation in the bud. If this happens, get the horse out of the feed, make a rational assessment of how much your horse may have eaten, and call your veterinarian and tell him or her that you have an emergency on your hands. Don't wait to see if the horse comes down with laminitis as a result of its eating binge. In an attempt to prevent laminitis, your veterinarian may consider using medications such as light mineral oil or activated charcoal (generally delivered through a tube passed through the horse's nose and into its stomach), to try to interfere with the metabolism and absorption of the feed. Any number of other drugs and treatments may be suggested (Figure 6) based largely on the experience and inclination of the veterinarian. The bottom line is that if you recognize circumstances that may make it more likely for your horse to develop laminitis and if you act promptly, you at least have a chance to keep things from getting started.

NUTRITIONAL STRATEGIES FOR THE TREATMENT OF LAMINITIS

With fat horses, weight reduction is the common goal of treatment. This at least helps to get some weight off the horse's sore feet, although such efforts are generally too late to have any significant

FIGURE 6

*Any number of drugs and treatments may
be suggested for the horse with laminitis.*

effect on the outcome of the disease. Do keep in mind that starving a fat horse is not a good way to treat laminitis—horses need proper nutrition to help them heal, so you should work with your veterinarian to develop a sensible weight-reduction program.

A number of hoof-related supplements may also be suggested. For example, the vitamin *biotin* or the amino acid *methionine* have been given as feed supplements to horses with laminitis because these substances are normally found in high quantities in the hoof. Similarly, all sorts of vitamin and mineral supplements have been advocated to "help." As a point in fact, nobody has any idea if providing these sorts of nutritional therapies is helpful, or even if they might have any *adverse* effects. So, make sure you consult with your veterinarian before taking any drastic dietary measures. You certainly want to do everything that you can, but you also don't need to waste money.

PHARMACOLOGIC STRATEGIES FOR THE TREATMENT OF LAMINITIS

Non-Steroidal Anti-Inflammatory Drugs (NSAIDs)

Almost without question, the most commonly used drugs in the treatment of laminitis are from a class of drugs that are nonsteroidal (that is, they don't have the chemical configuration that is typical of steroid drugs) and anti-inflammatory (that is, they help to reduce inflammation). There are a number of such drugs available to treat horses, as well as humans. In horses, the most common is undoubtedly *phenylbutazone* ("bute"), but others would include *flunixin meglumine* (Banamine®), *aspirin, ketoprofen* (Ketofen®), and several others.

Uncontrolled inflammation is usually not a good thing. While inflammation is a normal process that helps the horse's body isolate injurious agents and/or injured tissue, the complex series of events that characterize the inflammatory process can have all sorts of bad effects on the horse's foot. Most simply, inflammation causes local blood vessels to leak fluid. It also allows lots of inflammatory cells into the affected area. Leaky blood vessels cause swelling. In the case of the relatively rigid hoof, there's nowhere for the swelling to go. This can cause all sorts of problems. In an analogous situation, paramedics will cut a boot off the foot of a person with a sprained ankle, rather than let the leg swell inside the boot and risk increased pressure on the skin and resulting skin death. In some ways, it's a shame that this can't be done with a hoof. Furthermore, the inflammatory cells that are released into an inflamed area cause willy-nilly destruction of anything in the area; this should not go unchecked. Attempting to control the process is an obvious goal of treatment. Thus, the NSAIDs are often one of the first treatments suggested by the attending veterinarian.

NSAIDs have other potential benefits, as well. For one, they control pain. As you know, laminitis is painful, so there are some very humane reasons for using NSAIDs. That being said, some people do express concerns about the possibility of making a horse with damaged feet so pain-free as to allow it to walk around and cause further damage to the structures of the hoof. In general, it's probably not a bad idea to keep the amount of exercise that a horse with acute laminitis receives to a minimum, and perhaps it's even more important if the horse is receiving NSAIDs. Still, the drugs are just not all *that* potent when it comes to relieving pain (think about it, if you are really hurting, do you really expect a couple of aspirin to help?). Two, NSAIDs may also be useful in combating the effects of *bacterial toxins* released from the intestine. Finally, they're also "blood thinners," that is, they interfere with the action of blood clotting cells called platelets, and, as such, may help promote circulation to the damaged foot.

Horses with *acute* laminitis are often kept on NSAID therapy for at least two weeks. Horses with *chronic* laminitis that is difficult to control may be kept on the drugs almost indefinitely. There are complications associated with these drugs—most notably ulcers of the gastrointestinal tract and kidney problems—but the true incidence of these complications appears to be much less than the popular concern about them. NSAIDs are not "horse killers," and they can be an important part of the treatment regimen for a horse with laminitis.

Fentanyl Patches

While NSAIDs are almost certainly the most commonly employed method of pain relief for horses with laminitis, they do have some disadvantages, occasional side effects notwithstanding. One is that they have to be given regularly; it can be a bit of a nuisance to give a horse medication twice daily, everyday. Another is that they are not potent pain relievers, so when a horse is in severe pain, NSAIDs are not likely to do much to alleviate it.

One of the newer approaches to pain control is through the use of *fentanyl* patches. Fentanyl belongs to the group of medicines called *narcotic analgesics,* and can only be obtained from your veterinarian. Narcotic analgesics are often used to relieve severe pain in people, but less commonly in horses.

Fentanyl acts on the central nervous system. Applied to the forearm of horses, some veterinarians have used two 100μg/hr patches, a dose that is not associated with side effects, and have reported adequate pain control for 2 to 3 days. The patches have to be reapplied every two days or so, moving back and forth between left and right sides to give each side a break. The treated area is shaved and the patches are usually taped in place.

Not every horse responds to fentanyl patches, and some veterinarians are investigating higher doses (more patches) to see if pain relief can be improved without the side effects that occur with larger doses. Still, the patches may offer a reasonable and convenient alternative to NSAID therapy for some horses.

DMSO (Dimethylsulfoxide)

DMSO is a clear liquid that is most commonly used as an industrial solvent. However, it can also be used as a medication and given orally, or intravenously, as a diluted solution. The doses seem to vary among people who use it. One of the main purported benefits of DMSO is that it is supposed to run around and pick up compounds called *free oxygen radicals,* which occur with inflammation. Thus, at least theoretically, it might have some use in treating horses that have developed laminitis, even though there's really not any evidence that free oxygen radicals are involved with the development of most cases of the disease.

It's a tragedy when the beautiful damsel of theory is slain by the dragon of reality! In fact, there doesn't seem to be much except anecdotal evidence from some practitioners that DMSO alters the clinical course of any known disease, much less laminitis. Time may yet prove skeptics of DMSO wrong, but it's been used for everything from "soup to nuts" for at least 30 years, and still no one seems to be sure of what its benefits really are. The only thing that you can be sure about is that the horse that receives DMSO will smell really bad for a couple of days as it breathes the volatile chemical out of its system.

Drugs to "Improve" Circulation

Since many people believe that impaired blood circulation has something to do with the development of laminitis, a number of therapies may be suggested in an effort to improve or "increase" the circulation of blood to the hoof. As a practical matter, most of these therapies have not been shown to be effective, even though there may be a good theoretical basis for their use early in the onset of disease.

On the other hand, and again from a theoretical basis, you might want to be careful when using these agents *too* early in the course of laminitis, when the blood vessels are dilating anyway. However, after damage to the laminae has already occurred, you could argue that getting the circulation going is desirable. (These are the sorts of issues that you get into when you discuss theories, and obviously, it can be confusing; it may not even matter all that much!) Of course, individuals are going to select different therapies based on their own experiences, because there's not any treatment that's been shown to be "best." Some choices for drugs to affect circulation might be:

Pentoxyphylline – *Pentoxyphylline* (Trental®) is an interesting drug that is thought to allow red blood cells to change their shape more easily, which theoretically, would make it easier for them to move through blood vessels, especially ones that may be narrowed or otherwise damaged. Unfortunately, it hasn't been shown that this actually happens in practice. In normal horses given pentoxyphylline, changes in circulation have not been demonstrated.

Isoxsuprine – *Isoxsuprine* first became popular as a treatment for navicular syndrome. As such, it was theorized that the drug might improve blood circulation in the foot and relieve a theoretical cause of the condition. Unfortunately, more recent testing casts doubt on whether isoxsuprine can be of any real value at all for any condition, since it is rapidly removed from the bloodstream by the horse's liver. It has also been shown not to affect the flow of blood to the foot of healthy horses.

Acepromazine and its relatives – *Acepromazine* ("Ace") is a drug that's commonly used as a sedative. It also has a side-effect that causes dilation of the little blood vessels that occur in the outer portions of the horse, such as the limbs. It has been shown that acepromazine increases microscopic blood flow to the feet of healthy horses, but it hasn't been shown that those effects extend to the laminae. Unfortunately, if you use acepromazine, you have to give it very frequently, as many as four-to-six times a day. Be careful if you use this drug to treat a stallion because it has been associated with the subsequent inability of some horses to retract the penis.

Other, similar drugs to acepromazine, such as *chlorpromazine* and *promazine* have also been used to treat laminitis, although they are not commonly available. *Phenoxybenzamine,* another drug that improves blood flow through little blood vessels has also been tried. Although not commonly used, it was shown to be beneficial in the treatment of horses in the developmental stage of laminitis in one study.

Heparin – *Heparin* interferes with the ability of blood to clot. Theoretically, this could help keep blood flowing through the injured tissues of the foot. However, blood clots have not been shown to be a feature of the feet of horses affected with laminitis so it's reasonable to wonder how much it might help. Still, high doses of heparin did have some use as a preventive for laminitis in at least one study. But there's conflicting information out there, for example, it has been reported to be both effective and ineffective in preventing laminitis in horses with gastrointestinal problems.

Nitroglycerin – *Glyceryl trinitrate* is most commonly applied as a patch over the blood vessels running down the back of the pastern. It's certainly a potent dilator of blood vessels in humans, which is one reason why it's used in people with heart troubles. Nitroglycerin got popular as a treatment for laminitis in the mid-1990s, but its use seems to be waning. There are some pretty good reasons for this. If you consider that the nitroglycerin is most likely absorbed through the capillary network system of the skin, its next stop would be the venous side of the circulation. The venous circulation would take the drug to the heart and then it would circulate back around the body before it hit the arterial supply to the foot. This means that any drug is going to be really diluted before it gets to where it needs to be. Furthermore, since the drug stays in the system for a very short period of time, from a physiological and pharmacological point of view, giving the drug in this way is probably very unlikely to have the effect that you'd like. Lastly, and most importantly, research has shown that nitroglycerin patches don't work. Investigators using a black walnut laminitis model concluded that treatment with nitroglycerin after development of clinical signs of laminitis did not have a significant effect on blood flow to the laminae. Probably time to find another drug...

Other Circulatory Therapies

Cold therapy – *Cold therapy* has been used for ages to treat the feet of horses with laminitis; one old remedy was to stand a horse with sore feet in a stream. Recent experiments have shown that cold, which causes blood vessels to constrict, has a protective effect during the developmental stages of laminitis. Exposure of the laminae of the feet to laminitis trigger factors circulating in the blood may be limited by keeping the feet cold. Cooling the feet may be a way of closing the pipes that bring the laminitis trigger factors to the feet. In addition, enzymes, and the chemical reactions that they get involved with, don't work as well when it's really cold. Perhaps cooling the feet might be able to slow down the *proteinase enzymes* (see p. 16) that are at work during laminitis, as well.

The basis for suggesting that cooling or icing the horse's feet may be a preventive strategy worth trying during the developmental stage of laminitis was an experimental observation. Investigators in Australia who were studying laminitis noticed that only horses with hot feet, (those in which the blood vessels are dilated), developed the condition. Horses that were able to maintain cold feet, (those in which the blood vessels are constricted), were laminitis free. These observations were made during trials done in an effort to cause laminitis in a climate-controlled laboratory.

Cooling of the feet can be effectively achieved by packing the feet in crushed ice or by soaking them in cold ice water. Unfortunately, to be effective, cold therapy has to be applied continuously. This can be done with no ill effects on the hoof; experimentally, horses have stood in ice baths for 48 hours with no problems. In fact, such lengthy treatment time would appear to be almost mandatory when you consider that the underlying triggers for laminitis (whatever they are) are being produced continuously during disease episodes. Of course, keeping your own horse in an ice water bath requires an extraordinary amount of time and dedication, but it is possible, and may well be worth the effort.

Icing horses feet is safe. Horses don't seem to show any discomfort when their feet are close to zero (humans would whine incessantly in similar conditions). It's certainly *possible* to make the horse's feet cold; sophisticated measuring techniques have demonstrated a significant *reduction in blood flow* to iced feet. There's also evidence that icing helps bodies deal with problems related to toxic factors delivered in the blood. For example, humans undergoing cancer chemotherapy can prevent the loss of scalp hair if they are prepared to have their scalp cooled with iced water when the medication is at peak blood concentration.

The case for the safety of the technique (using a slurry of crushed ice) in normal horses has been established. The theory is there. However, if cold therapy prevents laminitis, it will work best during the developmental phase. There could be some theoretical benefit to stopping further damage if it slowed down the *proteinase enzyme* activity. But remember, it's still theory, and the true benefits of icing the feet of a horse with laminitis still need to be established.

61

Heat therapy – Yes, heat (the opposite of cold—you can see how impassioned debates can get going as to the "best" way to treat a horse with laminitis). The reason that you might consider applying heat is if you ascribe to the theory that the feet lack blood supply early in the disease. Heat is known to cause blood vessels to *dilate* (that's why your skin turns red in a hot shower) and hot water soaks have been advocated for the treatment of horses with laminitis. However, there's no good evidence that they work.

Thyroid Supplementation

In the early 1980s, as mentioned before, it was suggested that horses might develop laminitis because the function of their thyroid gland wasn't up to snuff. In people and dogs, *hypothyroidism* is associated with obesity; obese horses tend to get laminitis; therefore, perhaps obese horses had thyroid problems. In fact, true hypothyroidism is almost unheard of in horses and it does not cause laminitis.

Some people treat horses that have laminitis with thyroid hormone anyway. Thyroid hormone increases the metabolism—when treating horses that have laminitis, it's apparently hoped that by increasing the metabolic rate, the horses will lose weight more quickly. While this may or may not occur, the best thing is to use proper diet and exercise as a means for weight control in order to keep the horse from developing problems in the first place.

Antihistamines

It was once thought that histamine, a chemical that's released during the inflammatory process, was very important in the development of laminitis because of its constrictive effects on blood vessels. Accordingly, antihistamines were once very commonly given for the treatment of laminitis. However, a specific role for histamine has not been found and the use of these drugs in laminitis cases appears to be on the wane.

Nerve Blocks

Nerve blocks have a good theoretical basis for use in the treatment of laminitis because they block pain and cause blood pressure to drop. They may also help improve the local circulation of blood. However, completely relieving the pain in the foot of a horse with laminitis also makes the horse more likely to bear weight on it, thus potentially increasing the damage and making a bad situation worse. Long-term use of nerve blocks to treat horses with laminitis is unwise.

Bleeding
Arguably, the oldest treatment in medicine is to release some blood from a sick patient (phlebotomy). It's certainly been used to treat laminitis for centuries and some people even practice it today. There's no known reason why such a treatment should be effective.

"Alternative" Remedies
One of the sure signs that a disease is not curable is when many different treatment approaches are offered. This mostly reflects the fact that, in many cases, people are grasping at straws. In areas where sure-fire solutions do exist, there is no "alternative." For example, one doesn't see "alternative" fields of airplane engineering, or bridge building.

However, in the treatment of laminitis, since no treatment is sure to cure the condition, there are any number of alternative options. For example, one published approach to therapy states that since *phenylbutazone* causes changes in the ability of the intestinal wall to let things pass through it, and the unidentified factors that are at the root of laminitis may come via the gastrointestinal tract, by not using phenylbutazone, or by *removing it* (and, presumably, similar drugs) if they've been given, laminitis becomes more responsive to treatment. Under such a scenario, nutritional management and the application of other modalities, including acupuncture and Chinese and Western herbs, are said to be useful in "healing" the intestinal wall and "completing" the healing process.

Of course, such a scenario overlooks the fact that throughout most of history, phenylbutazone *wasn't* available and horses still developed, and failed to respond to, treatment for laminitis. And, in general, the only thing that alternative approaches for the treatment of laminitis have in common is that none of them have ever been demonstrated to be effective in good research studies. In fact, many of them have no plausible biological or scientific rationale. The bottom line is that there is currently no good evidence to support the use of treatments such as acupuncture or homeopathy in tending to horses with laminitis. However, they are unlikely to hurt your horse, particularly if they are used in conjunction with therapies that have a more plausible rationale.

In cases of laminitis that are self-limiting, that is, in cases in which the horse is going to get better anyway, any therapy will be thought of as having been successful. That's one of the reasons why therapy for the condition is so confusing—there is not enough information currently available to suggest that one treatment approach is the "best."

Many horses that develop *acute* laminitis go on to recover completely. However, even the mildest cases require careful care and close observation. If the bone stays in place and the clinical signs of laminitis do not return for several days after treatment has stopped, it is probably okay to carefully and slowly begin returning the horse to its normal function. On the other hand, if the bone has moved inside of the hoof, the chance for full recovery is more guarded, and medication and meticulous foot care may be required for life. It is certainly possible for there to be some *mild* rotation of the bone of the foot and for the horse to still make a complete recovery and to remain sound indefinitely. But, the more the coffin bone moves inside the hoof, the worse things are likely to be. In spite of prompt and proper medical therapy, horses that have significant destruction of the laminar tissues that keep the foot together may *never* make a complete recovery and may experience repeated episodes of foot pain. Ultimately, more than any particular therapy given, the likelihood of an individual horse making a full recovery is directly related to the severity of the damage to the tissues connecting the coffin bone to the hoof.

Dr. Nat Messer is a 1971 graduate of Colorado State University's College of Veterinary Medicine. He spent 12 years in private practice in Colorado, followed by 18 years academic clinical practice at Colorado State University and now at the University of Missouri. Dr. Messer is currently an Associate Professor of Equine Medicine and Surgery at the University of Missouri and a diplomate of the American Board of Veterinary Practitioners, certified in equine practice.

6

Foot Care

With Steve O'Grady, DVM

The "best" way to take care of the foot of a horse that suffers from laminitis is still the source of a tremendous amount of controversy. In fact, there probably is not a "best" way; it most likely varies from horse to horse. That said, it's almost a given that advances in both the technology available for the care of the horse's foot, as well as a greater understanding of the forces that come to play on the horse's foot have allowed for some horses to be saved—even to return to partial or full athletic function—that might otherwise have been lost in previous years. Still, in spite of some successes, there are far too many failures. Keep in mind that the outcomes are directly related to the extent of the damage suffered at the beginning of the problem. Even the best veterinarian and farrier team will not be able to save every horse.

FOOT CARE IN THE ACUTE STAGE OF LAMINITIS

When you recognize that a horse has laminitis, the first thing that you need to try to do is to attempt to relieve the associated pain and inflammation. This is where medical therapies come in. Still, you certainly can't count on medications to solve all of the horse's problems. The foot itself generally needs some direct help. There are several ways that veterinarians and farriers may try to provide it.

One common recommendation is to try to support the horse's foot. Placing the horse in a stall with a floor bedded deeply with sand, for example, can do this. Sand, or other similar material, provides a lot of cushion and support for the bottom of the horse's foot, effectively filling the contours of the ground surface of the hoof and pushing up into it (very much like the diagrams that you've seen for

FIGURE 7

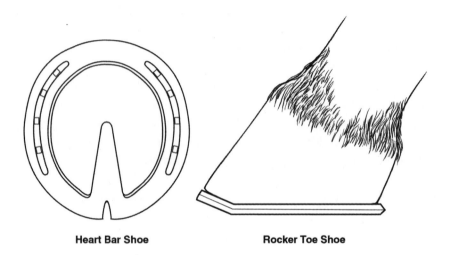

Heart Bar Shoe   Rocker Toe Shoe

The heart bar shoe and rocker toe shoe. Two of the many shoes tried for the treatment of laminitis.

a good supporting mattress). You also hope that this counter pressure might also tend to resist the movement of the coffin bone in the hoof, but no one really knows if that happens or not.

Another common strategy is to try to change the way that the foot is bearing weight. Generally, a horse with laminitis is most sore in its sole. So, a horse may benefit from you trying to apply pressure somewhere else on the foot besides the sole. The largest and best area to try to get this done is on the frog of the hoof: the V-shaped area that you see in the center of the hoof when you pick it up. Frog support can be provided in any number of ways, from a simple roll of gauze taped in place, to more esoteric methods such as a heart bar shoe (Figure 7), which is constructed with a metal tongue over the frog. Some people assert that frog pressure also tends to counteract the tendency of the coffin bone to rotate downward, but there's really little evidence that this actually happens. Still, even if it doesn't help keep the bone from moving, frog pressure can provide pain relief for some horses.

Another approach is to try to cushion and support the entire foot. As such, thick industrial Styrofoam has become very popular for the management of some horses with laminitis. Styrofoam board is used in construction and you can get it at just about any store that sells building materials. It's simple to stand the horse on a piece of the material so that an impression of the hoof is made in it. Then, you cut out the foot-shaped piece of the Styrofoam and tape it in place. In this way, Styrofoam acts in much the same way as deep sand, but it's a lot more portable.

Shoeing in the Acute Phase

It's not at all clear when you should start trying to shoe a horse with laminitis. Still, based on current knowledge, it seems reasonable to suggest that you might first consider shoeing when the horse has become somewhat more comfortable, is on minimal amounts of medication, and the coffin bone has stopped moving inside the hoof. Otherwise, you might find yourself having to change the shoe all of the time, and all of that pulling and pounding might just make the horse more sore and uncomfortable.

It's reasonable to ask, "What's the purpose of shoeing a horse with *acute-phase* laminitis anyway?" After all, what with the variety of ways that you can apply a shoe, the myriad of shoe and pad combinations, and the fact that no one technique will work in every horse, it's pretty hard to say with assurance that any one way is better than another. It's tempting sometimes to just throw up your hands in confusion. But don't despair, in general, there are five specific principles that people try to follow when they attempt to get a horse with acute laminitis back on the road to recovery.

1. Protect the foot. This one's pretty simple. You're trying to add some protection, to keep the foot from contacting the ground. Shoes are one way to get this done. Indeed, it's the same reason that you shoe *any* horse.

2. Ease breakover. *Breakover* is a term that refers to a normal movement of the horse's foot during the stride. The hoof "breaks over" the toe just before it comes up off the ground. Breakover is painful for the horse with laminitis because it stresses the compromised laminae. To the extent that you can make it easier for the horse to accomplish this movement, you can make it more comfortable.

   To minimize the stress of breakover, the front of the shoe should be placed just in front of the *apex* or *leading border of the coffin bone.* This point is determined by taking X rays of the foot, using markers such as a thumbtack in the frog (don't worry, the horse won't feel it). The X rays then show where the thumbtack is in relation to the coffin bone. This picture helps orient your veterinarian and farrier to the location of the underlying bone so that the shoe can be properly placed.

3. Relieve the pressure from the front of the hoof wall (your veterinarian will call it the dorsal hoof wall). Always keep in mind that the laminae of the hoof are where all of the action in laminitis is going on. So, you want to do whatever you can to try to keep from putting extra pressure on the laminae. Shoeing methods that transfer weight to the hoof wall, especially in the toe area, most likely increase pressure on the compromised laminae. This may worsen or prolong the problem and be much worse than no shoe at all.

People have tried things like turning the shoe around and putting it on backward, leaving the toe completely open, so as to avoid putting extra pressure on the hoof wall. This may or may not work, but it might even *increase* pressure on the toe if, by opening up the toe, it allows for more contact with the ground.

4. Prevent sole pressure. Since the sole is the area that's most sore, it makes sense to not put extra pressure on it. So, you should avoid any sort of shoe or shoe and pad combination that pushes up against the sole of a sore-footed horse. You might even think about putting a bevel on the shoe—grind the shoe so that it slants away from the sole—to prevent too much sole contact.

5. Provide posterior support. In the horse with laminitis, the frog and heels of the foot can become useful weight-bearing structures. This can be done in any number of ways, from using specially designed pads that fit under the shoe, to the previously mentioned heart bar shoe. It's also important to make sure that the shoe is fitted well behind the horse's heel, to maximize the amount of surface on which the horse can bear weight.

If, by shoeing, you can't achieve the five points mentioned above, you're probably better off leaving the horse barefoot. In fact, some horses seem to be much more comfortable when they're *not* wearing a shoe. Maybe pounding nails into an already sore foot makes it more uncomfortable. Maybe clinching the nails down to make sure that the shoe stays on increases the overall discomfort. At any rate, keep in mind that since your horse is hurting, you'll be tempted to not just stand there, but will want to do something. However, sometimes things work better if you leave them alone (that is, don't just do something, stand there).

CARE OF THE FOOT OF THE HORSE WITH CHRONIC LAMINITIS

It's not completely clear at what exact point a horse should be called *"chronic."* Most people would assert that a horse becomes chronic when the first X-ray changes are seen. That is, a horse can't really be said to have long-term laminitis until there's some sort of change such as a rotation and/or displacement of the bone within the hoof. If your horse gets over an episode of laminitis, you might not have to do a whole lot to its feet, but once changes have occurred, you'll have much to consider.

SHOES AND PADS

There are countless types of shoes and pads that have been used in the treatment of horses with laminitis. This farrier's smorgasbord reflects the fact that there's no such thing as a "best" shoe or pad (or combination thereof) and that no single shoe or pad will help every horse. This is not necessarily a bad thing, because it means that you'll have a good number of options to try. Here's a short list of what's available:

1. The heart bar shoe (Figure 7). Heart bar shoes were originally used way back in the nineteenth century, and they had resurgence in popularity for the treatment of laminitis in the 1970s. The shoe has a welded metal "tongue" that puts pressure up against the frog. This may help transfer some of the weight of the horse from the sole to the frog. The frog isn't normally involved in a significant amount of weight bearing but it's quite useful to get it working for the horse with laminitis. Some eager proponents of the shoe have gone so far as to say that upward pressure from the welded-on bar resists the tendency of the coffin bone to rotate inside the hoof, but this really hasn't been shown to be the case. Factory-made heart bar shoes come in a number of sizes and can be shaped to fit the foot of just about any horse. Of course, a good farrier can make one, as well. The shoes also come in plastic and these can be glued onto the foot.

    It takes a good bit of skill to properly apply a heart bar shoe. On the one hand, if the metal tongue is just hanging out there in the air, and doesn't provide any frog pressure, it's not going to do much good at all. On the other hand, if the bar causes too much pressure on the bottom of the frog, it can cause the horse a considerable amount of pain.

It's certainly *not* the case that a heart bar shoe is the only way, or even the preferred way, to treat a horse with laminitis, so don't go thinking that your horse with laminitis hasn't been treated properly if it's not wearing one. But the heart bar shoe has received a lot of press over the years. Now you know about it, too.

2. The egg bar/heart bar shoe (full support shoe). This shoe, is, as the name implies, shaped like an egg, with a bar of metal welded in to fit over the frog. It provides full support to the hoof.

3. An egg bar shoe with a square, rolled, or rockered toe (Figure 7). Farriers can take the front of a horseshoe and change its shape by heating it and forging it with a hammer. By squaring or rolling up the toe of the shoe, they try to help the horse breakover its toe with less effort.

4. A reverse shoe, which leaves the toe open. This is done in an effort to get all of the pressure off the toe and the front of the hoof wall. Unfortunately, while it looks good in theory, turning a shoe around doesn't always work very well. In some horses, when they breakover, *more* pressure may be placed on the now unprotected toe. Reverse shoes are often used with a frog-support pad.

5. An *open* shoe with extended heels, using a heel insert such as a frog support pad.

6. A *wedge* shoe. This may be used in an effort to lift up the foot and relieve pressure on the deep digital flexor tendon (see below).

7. Various types of *glue-on shoes* and modifications thereof. These shoes offer an advantage by not requiring nails to be placed in the sore foot. They are used very successfully by some veterinarians and farriers.

8. *Any* type of shoe that puts a lot of metal under the horse's foot, in an attempt to increase its weight-bearing surface.

THE THREE MOST IMPORTANT FORCES ON THE FOOT OF THE HORSE WITH LAMINITIS

Even though there are many different types of shoes and pads, all shoeing methods for laminitis should try to address the same basic principles. That is, they should all be applied with the idea of resisting the three main forces that come to play on the foot of a horse with laminitis.

Weight

The first force is the load that the horse's weight puts on its compromised feet. The various types of shoes try to counteract this tremendous force by providing some type of support to the bottom of the foot. There is any number of ways that shoes and pads attempt to do this. For example, heart bar shoes may help transfer much of the weight-bearing function from the hoof wall to the frog. Weight transfer can also be accomplished through the use of wide-web steel or aluminum shoes or various bar shoes, which increase the amount of weight-bearing metal on the ground. Frog-support pads attempt to do the same as heart bar shoes. *Polysiloxane* impression material *(elastomere putty)* fills the foot with a pliable support material. All of these methods use the posterior portion of the foot, that is, the heels, bars and frog, in an effort to help support the foot and get weight off of the sore sole.

The Pull of the Deep Digital Flexor Tendon

The second major force to consider is the pull on the coffin bone that normally comes from the deep digital flexor tendon tending to pull the toe down toward the ground. Veterinarians and farriers may try to counteract this force in two ways.

First, they may suggest lengthening the shoe to give the foot a broader base of support. For example, the heel of the shoe can be extended beyond the heel of the foot. This does a couple of things. It may help decrease the force exerted by the deep flexor tendon. And, a longer shoe means that the weight of the horse gets transferred farther back on the foot, and away from the toe. You might be able to get an indication that lengthening the shoe is helping as you follow the progress of the foot with X rays (as you should). You hope to see that the sole gets thicker over time, which means that the tissue is healthy and growing. If the tissue is healthy and growing, it most likely means that the shoes have taken some of the load and pressure off of the toe.

Second, veterinarians and farriers may try to manipulate the hoof angle to make it steeper. That is, they may try to raise the heel of the hoof up off the ground in an effort to reduce the tension in the deep flexor tendon. By lifting the heel, and thereby increasing the geometric angle that is formed where the front of the hoof wall intersects the ground, you're basically trying to shorten the distance that the tendon travels. You're trying to put some "slack" in the system. To some extent, manipulating the hoof angle can be done with trimming, but other means of raising the heels, such as wedge pads, are commonly used. This elevation can be quite dramatic; some veterinarians have advocated raising the heels by 15 degrees or more!

The Bending Force at Breakover
The third, and final important force placed on the compromised tissues of the horse with laminitis is a bending force that occurs on the front of the hoof wall itself. This force occurs normally during the stride, when the horse's foot breaks over the toe, just before it comes up off of the ground. The force tends to bend the hoof wall backward, which means that pressure may be applied right over the top of the sore laminae. Veterinarians and farriers try to manage this force by moving the breakover point closer to the tip of the coffin bone (in order to do this effectively, you have to use X rays to see where the tip of the coffin bone really is). They may also try to help improve breakover by doing such things as squaring the toe of the shoe or using a rolled or rocker-toe shoe. The hope is that by rolling up the front of the shoe, there will be less resistance to breakover. This is also the idea behind open-toed shoes (or just putting a regular shoe on backward). But, whatever technique (or combination of techniques) is used to try to influence breakover, they're all being done in an effort to reduce the pressure on the damaged laminae of the hoof.

X RAYS (RADIOGRAPHS)
Over time, your veterinarian will most likely want to try to get the horse's hoof trimmed to get things back in normal alignment. You are simply not going to have a lot of success shoeing the horse with

laminitis if you don't get good X rays taken from time to time. Periodic radiographs are used as a baseline to follow how things are progressing. They allow the veterinarian and farrier to look at the bone of the foot and determine the degree of rotation and/or movement toward the ground that has taken place; the current position of the bone; and the thickness of the sole of the foot, (which is important to keep the coffin bone from hitting the ground). X rays also let you see such things as the laminae separating from the hoof, or areas where infections or abscesses may have developed.

Equally important, X rays should be used as a guide for trimming and shoeing the foot. When the X rays are being taken, markers placed on the foot can help your veterinarian figure out where things are and where the shoes should best be applied. For example, wires placed on the front of the hoof wall and on the bottom of the foot allow the angle to be more easily determined. As mentioned previously, a thumbtack in the frog can help the veterinarian and farrier locate the bone of the foot relative to the frog (see p. 68).

Measurements made from the X rays also can help determine the most accurate point of breakover at which to fit the shoe. They can help determine the amount of heel that must be removed and the point on the ground surface of the foot where the trimming should begin in an effort to get the bone back into a more normal position.

Measurements of the degree of flexion in the coffin joint should also be made. A horse with *chronic* laminitis, in which there is significant rotation of the coffin bone and a high heel, may also have a noticeably flexed coffin joint. This is particularly obvious when the horse bears weight. A perpetually flexed joint can be very uncomfortable for a horse and is something that needs to be addressed with shoeing.

In severe cases of laminitis, those with unrelenting pain and marked rotation of the coffin bone, some veterinarians may recommend surgeries such as a *desmotomy* of the *accessory (inferior check) ligament* or a *deep flexor tenotomy.* In these cases, radiographs are mandatory to allow you to attempt to return the foot to a more normal alignment (see Chapter 7).

APPLYING THE SHOE
Shoeing the horse with laminitis should always be a combined

effort between a veterinarian and a farrier. Input from both parties is necessary for a successful outcome. You need to keep in mind the three goals: 1) support the foot, 2) decrease the pull of the deep digital flexor tendon, and 3) ease breakover. In addition, when your farrier applies the shoe, ask him or her to make sure that they don't put any additional pressure on the horse's sole; one way to avoid this is to use a grinder to make the surface of the shoe concave where it comes in contact with the foot.

Support the Foot

If possible, the length of the shoe should extend beyond the heel of the foot to a perpendicular line drawn from the hairline at the bulbs of the heel to the ground. The shoe can be lengthened or shortened if necessary, depending on the conformation of the foot. You can add additional support to the foot by using composites such as dental impression materials, which form a firm, resilient material that molds to the heel area of the foot. By using this sort of material, which is usually placed between a wedge pad and the foot, all of the structures in the heel area that are capable of providing support can be used, and uniform pressure can be applied. In selected cases, you can even fill the entire surface of the sole with the impression material.

Decrease the Pull of the Deep Digital Flexor Tendon

A properly fitted shoe not only provides support to the foot and shifts weight bearing, it should also decrease the pull of the deep digital flexor tendon. To help with this, wedge inserts can be placed between the shoe and the foot. The proper amount of heel elevation can also be determined from the X rays. Ideally, these wedges will be removed over time.

Ease Breakover

The inside edge of the shoe should be placed at or just in front of the coffin bone. The precise spot can be determined from X rays.

SPECIAL CONSIDERATIONS

In the most severe cases of laminitis, the surface of the sole of the foot may become compromised and require treatment. In these cases, an aluminum treatment plate can be easily attached to the shoe. This allows access to the foot for treatment, but also provides protection to the damaged tissues. If there are infections in the hoof wall (known as "seedy toe"), it may be necessary to remove the hoof wall overlying the infection. These special cases require extra care and that the professionals involved work together.

Success in shoeing the horse with laminitis comes from careful attention to the horse and a close adherence to the basic principles of shoeing. Successful treatment is challenging and it usually entails long-term maintenance, and you can never get too comfortable with your results. Regardless of the types of shoes, pads, or trimming techniques that you use, it is important for you to ask your veterinarian and farrier to consider the biomechanical forces exerted on the foot. If you can do that, and if you can get them to work as a team, you'll make sure that you have the best chance of a successful outcome.

Dr. Steve O'Grady was a professional farrier for 10 years prior to obtaining his degree in veterinary medicine from the University of Cape Town, South Africa. After graduating from veterinary school, Dr. O'Grady completed an internship in Cape Town. He then moved to Charlottesville, Virginia, where he worked in private practice for 5 years before opening his own practice. He now works both in Virginia and South Africa, with a large portion of the practice devoted to care of the equine foot. In addition, Dr. O'Grady has published numerous articles and lectured extensively on equine foot problems.

7

Surgery

For most horses, laminitis is a medical problem. That is, in most cases, prompt, aggressive, proper care and medication will be the strategies that are most likely to be effective in resolving the problem. Unfortunately, in spite of everyone's best efforts, there are some horses that do not respond to proper and prompt care. For these horses, surgery may become an option.

Surgery should be entered into with a good deal of thought. In general, once you've decided to do surgery on a horse with laminitis, you're in for a lot of aftercare, so you should not consider it if you cannot pay attention to your horse afterward. Also, don't go into surgery with unrealistic expectations. As with just about everything else associated with the disease, surgery does not always meet with success. The decision to do surgery often ends up being one in which you pay your money and take your chances.

DEBRIDEMENT (REMOVAL) OF DAMAGED TISSUE

The most straightforward surgeries for laminitis involve the removal of dead or damaged tissue. Procedures to do this are often done with hoof tools and usually don't require any anesthesia or travel to a vet hospital. For example, if there's some dead sole material, or if there is some hoof wall that has separated, your veterinarian and/or farrier may feel that it's prudent to pare away some of it with a hoof knife, nippers, or even a motorized tool, so that it can be replaced by healthy tissue.

Occasionally, removal of damaged tissue can become much more involved. When infection goes deep into the tissues of the foot, or when portions of structures such as the coffin bone become devitalized, it may be necessary to consider more aggressive tissue removal under anesthesia. Horses that have involvement of these deeper tissues are generally serious, *chronic* cases and, in many cases, the procedure may be undertaken in an effort to salvage the horse's life. Depending on the value of the horse and the commitment of the owner, it might be worth a try, but go into something like this with your eyes open and your fingers crossed. If a horse has involvement of deeper structures, the odds are not in its favor.

DORSAL HOOF WALL RESECTION

In the mid 1980s, it was proposed that if you took the front part (the dorsal part, in medical terms) of the wall off of the hoof, you might be able to help relieve some of the pain and inflammation caused by laminitis. The theory was that by removing hoof wall, there wouldn't be so much pressure on the top of the coffin bone. Some people felt that there also wouldn't be any pressure on the coronary band, which would theoretically allow new hoof to grow down more normally. Without the hoof wall to keep a lid on the hoof, some further theorized that you might be able to more easily put the bone of the hoof back into its normal position.

This technique was usually done in association with the application of a heart bar shoe. For a while, it was part of the routine treatment of laminitis by some veterinarians and farriers. However, as with other aspects of laminitis, there was a bunch of beautiful theory, but ultimately precious little to support it. The initial reports of success have not panned out, and over time, it seems that veterinarians and farriers are employing this technique less and less.

That's not to say that the surgery has absolutely no value at all. Some cases—those where there have been complications such as separation of the hoof wall, with infection and debris between the hoof layers—may benefit from removal of portions of the wall of the hoof. In fact, these cases usually can't be successfully treated by any other means. Getting rid of dead tissue encourages healing; so, in selected cases, careful removal of hoof wall and associated dead tissue (occasionally, even some bone) would seem to be at least *potentially* helpful. Unfortunately, nobody really seems to keep good track of such cases, so they tend to be handled on a case-by-case basis, and "success" probably depends on a whole lot of factors, including the skill of the people involved. Regardless, keep in mind that if you are considering having part of a horse's hoof wall removed, it's going to mean a lot of time, expense, and aftercare.

ACCESSORY (INFERIOR CHECK) LIGAMENT DESMOTOMY

The *accessory ligament* of the deep flexor tendon (more commonly called the *inferior check ligament*) is a short strap of tissue that goes between the horse's suspensory ligament and the deep flexor tendon, in the back of the cannon bone region where all of the big tendons and ligaments run. The check ligament gets its name because it is thought that it "checks" excessive movement of the deep flexor tendon. Noted equine anatomist, James Rooney, DVM, feels that check ligaments help dampen the vibration associated with dynamic impact loading. During standing and quiet walking, he thinks that these ligaments are energy-conserving devices that sustain the body's weight without "expensive" muscle work— movement that might otherwise cause the tendon to undergo excessive strain.

Some horses with *chronic* laminitis may have a flexed coffin joint (when viewed on X rays), a distorted hoof capsule, unequal hoof wall growth, or they may not grow very much sole. Some veterinarians have advocated cutting the check ligament when the coffin bone has stopped moving (as seen on X rays), there is no obvious discomfort, and minimal medication is required, but where the horse is still getting around with a short "stilted" gait. Cutting the check ligament is done in an effort to release some of the tension on the deep flexor tendon, but, obviously, the release of tension is not to the same degree as when the tendon itself is cut (see below). It is a less aggressive surgery than is cutting the tendon of the deep digital flexor muscle, however, the release of tension may be enough to allow a skilled veterinarian or farrier to realign the bottom of the coffin bone so that it is parallel to the ground (some people refer to this as "derotation"—see p. 82).

After the check ligament is cut, good aftercare is important. Routine wound care and bandage changing, as directed by your veterinarian, will be an obligation for at least two weeks after surgery. In addition, walking the horse is generally recommended in an effort to help keep the tissues from contracting.

DEEP DIGITAL FLEXOR TENOTOMY

Sometimes, in spite of aggressive and completely appropriate care, things continue to deteriorate. The bone of the foot may continue to rotate or sink inside the hoof. Pain may become unrelenting, and even high levels of medication may not control it. This is when you might want to consider cutting the deep digital flexor tendon.

The reason to do such a drastic procedure is really quite simple. What you're trying to do is stop the pull of the deep flexor muscle on the coffin bone. As you recall, the deep flexor tendon pulls the toe up and back when its muscle contracts, and its force is normally directed in such a way that it pulls the tip of the coffin bone down (of course, normally this pull is resisted by the laminar connections inside the hoof). If you cut the deep flexor tendon, you make it impossible for the muscle to pull on the bone and, you hope, you can slow down or stop its rotation.

The surgery itself is quite simple. It's usually done in a standing, sedated horse, under local anesthesia, at the level of the mid-cannon bone. The surgical wound usually doesn't even require much after-care—just routine wound care and bandaging, as directed by the veterinarian who did the surgery.

Deep digital flexor tenotomies are often combined with a *derotation* (described in the following section) to immediately return the coffin bone to a more normal position. Obviously, you need a skilled veterinarian and farrier working together to pull this sort of thing off. After surgery, you should expect to rest your horse in a stall for at least six to eight weeks. (And, keep your fingers crossed.)

Some horses show remarkable improvement immediately after deep flexor tenotomy and recover to lead completely normal—in some cases even athletic—lives. However, other horses may not show any improvement at all. In fact, the single study that did a long-term follow-up on horses treated with this surgery did not show it to be ultimately helpful. Still, when all other options have been exhausted, or the horse is not responding to treatment, a deep digital flexor tenotomy may offer a final chance.

When to do the procedure is still a bit of a question. Like any surgery, you don't want it to be a last-ditch effort just before you throw in the towel. Indeed, may veterinarians feel that the problem with lack of success in many of these procedures is that it isn't done

FIGURE 8

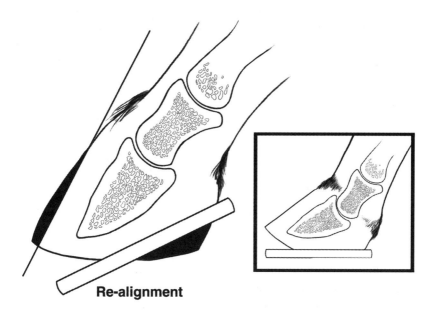

Re-alignment

Realigning the coffin bone means trying to reestablish the normal hoof angle. To do that, it's necessary to cut back the toe and remove heel (shaded areas). The resulting foot may look a little funny, but with luck, it will be on its way to growing out normally.

soon enough. So, if your horse with a bad case of laminitis isn't responding after a few weeks, and you have the time and inclination to try to do something heroic, ask your veterinarian about this surgery. There are living horses performing athletic activities demonstrating that it's possible to come back from even a dramatic procedure such as this.

REALIGNMENT OF THE COFFIN BONE (DEROTATION)

Whenever you cut the check ligament or the deep flexor tendon, most people feel that it is a good idea to try to realign the coffin bone so that it's parallel to the ground at the same time. This process has also been called "derotation." Indeed, some people feel that a failure to do this may be the main reason for many of the failures of the surgeries that are reported in the veterinary literature. It doesn't make much sense to try to release the tension on the deep flexor tendon and then leave the coffin bone pointing down into the ground. If you're going to go to all of the trouble to do the surgery, you probably don't want to leave this additional therapeutic stone unturned.

When realigning the coffin bone, it's essential that you use X rays as a guide. Lines drawn on the X rays serve as a guide for how much tissue needs to be removed. One line goes through the portion of the heel that is to be removed; that line stays parallel to the bottom of the coffin bone along its entire distance and will be the new ground surface that supports the foot. The shoe goes under this line. Another line is drawn parallel to the upper (dorsal) surface of the coffin bone; this line will intersect the line drawn on the ground surface (Figure 8). Using the X-ray guide, the foot is trimmed appropriately, a shoe is placed, and, voila, the horse's foot is realigned!

Horses that have had their coffin bone realigned usually end up with a good part of the toe having to be chopped off. The foot will look strange for a while until the toe grows out. Following trimming, a shoe may be applied with nails, but some veterinarians are reporting more success using glue-on shoes, which have the advantage of keeping nails out of the feet.

One other thing about placing the shoe is important. When a horse's deep digital flexor tendon is cut, the toe has a tendency to lift up into the air. That's because, due to the surgery, a very important supporting structure has been eliminated. To counteract this tendency, the shoe should be fitted so that it extends one-half inch or so behind the heel.

Unfortunately, even surgical techniques appear to be of limited value when the whole coffin bone is sinking. If all of the laminar connections in the horse's foot have let go, there's really not much that can be done to keep the horse's weight from driving the bone of the foot right into the ground. However, even in these severe cases that are simply not responding to your best efforts to provide appropriate care, a combination of surgery and good farriery may be able to turn things around. You have to decide if it's worth the risk, money, and effort.

8

Prevention

with Christine King, BVSc, MACVSc (eq med),
MVetClinStud

The old adage that an ounce of prevention is worth a pound of cure is especially true of laminitis. Since it's a given that many horses that develop laminitis will be permanently affected—and some will not recover at all—it makes a lot of sense to do whatever you can to keep your horse from developing the problem in the first place. Of course, that's not always possible, but there are some management procedures to keep in mind to decrease the likelihood that you'll ever have to deal with laminitis.

DIET

In general, the easiest thing for you to control is your horse's diet. The dietary key to preventing laminitis is to limit its intake of easily digestible carbohydrates.

Pasture

As you know, even though fresh grass pasture is the most "natural" and least expensive way to feed horses, and good quality pasture can be a complete nutritional diet for horses, it can also be a potential dietary land mine. Access to lush pasture has been identified as the underlying cause of almost 50 percent of all cases of laminitis.

Fortunately a few easy solutions to pasture-related problems exist. You can just limit your horse's access to the pasture in the first place; you can keep the horse off pasture until the grass is more mature; or simply limit grazing time to a couple of hours each day. Since, on sunny days, the levels of the fructan carbohydrates (see p. 39) peak around the middle of the day, you may want your turnout hours to be in the late afternoon or first thing in the morning. You can also put a grazing muzzle on the horse.

Another solution: some people simply offer moderately good quality grass hay in the pasture. By doing this, you're not trying to give the horse more feed—you're trying to give it more fiber. Horses have a requirement for the relatively indigestible portions of their feed (hay stems and such) and it's been shown that horses will try to fill their requirement for dietary fiber if it's not available. If the pasture grass is lush and full of water and fructans, it's also likely to be low in fiber. Even though there's lots of green stuff for them to eat in the field, horses will still consume hay if there's not enough fiber in the grass. Conversely, if the pasture is meeting their fiber and energy needs they will ignore the hay. Still, keep in mind that although putting hay out may be easier than taking the horses in and out of the pasture, it may not be as foolproof a way of management, and most especially not for horses or ponies that have developed laminitis when on grass pasture previously.

Grains

To help prevent laminitis, you can also easily limit your horse's access to whole or processed grains. Grain comes in many forms, whether rolled, crimped, or as any number of processed or extruded feeds. (Each bag of feed has a label attached to it, and it's a great source of information.)

For horses to live and breathe, they need energy to run their bodies. That energy, which is given to them in their feed, is measured in calories. Grains are an easily available source of calories, and some horses, particularly those that are exercising heavily or growing quickly, may need lots of calories. Still, people tend to feed their horse grain for a variety of other reasons, including sentimental ones, such as trying to make the horse happy. But the fact is that most horses don't need grain and are better off without it.

Grains, like lush pasture, are high in easily digestible carbohydrates and very low in fiber. This is the perfect dietary prescription for developing laminitis. The more you push grain into your horse, the more likely it is to develop a problem. And it's not just laminitis; high-grain diets have been associated with all sorts of other problems, such as colic, exercise-related muscle problems ("tying up"), obesity, and even bad behavior.

It usually takes several pounds of grain at one time to cause laminitis in an adult horse, but it may require much less in ponies. Avoiding grain is the best way to prevent problems, and most horses and ponies don't need it. However, some horses do require extra calories—those that exercise hard and regularly, for example—and grain is a good source of them. If you are going to feed grain, you shouldn't feed more than about 5 pounds per feeding.

You might also want to keep in mind that your horse may not necessarily think that restricting its grain is a great idea; make sure that you keep the area in which you keep your grain secure with latches on doors and tops on storage bins. Just be careful and sensible; grain doesn't kill horses, and it is certainly possible to safely give it to a horse.

For the horse that does need extra calories, you might consider increasing the amount of hay or hay products in its diet, or by adding fat to it. Fat sources such as rice bran can be fed alone, and liquid fat in the form of vegetable oils usually needs to be added to some sort of carrier, for instance, wheat bran or chopped hay. Of course, don't go about making any dietary changes for your horse without first discussing them with your veterinarian.

Body Condition

Fat horses are unhealthy horses. Fat horses can't tolerate exercise as well as horses in good condition: Fat acts like a layer of insulation around the horse. Being so well insulated, the fat horse can't get rid of body heat that's generated during exercise, which means that a fat horse can overheat and fail to perform well (or stop performing altogether). Chronic obesity is definitely associated with laminitis, as well as with the occurrence of fat tumors (lipomas) in the abdomen. (Lipomas can twist around the intestine and cause colic that will need surgery to correct.) In addition, heavy body weight puts an increased load on the feet. Furthermore, fat horses tend to be more inactive than horses kept in good condition, and inactivity itself is a risk factor for laminitis. Finally, there's the odd problem of obesity-associated laminitis that's discussed in Chapter Four.

Ideal body condition is simple. You can feel it. You should easily feel the ribs of a horse that's in good condition, although you won't be able to see them. If your horse has a big crest on its neck, or if you need to get a running start in order to get up enough momentum behind the push needed to feel your horse's ribs, then you need to get some weight off of your horse.

If your horse is too fat, it's generally your fault. After all, horses love to eat and they can only eat what you feed them or allow them to eat. The fact that horses in the wild eat for as long as 17 hours in a 24-hour period is irrelevant because those horses tend to eat low-calorie grasses and walk around all day; this is a perfect prescription for staying fit and trim. Their domesticated, overweight cousins sit in stalls all day, eating higher calorie feeds and buckets of horse treats. Horses kept in stalls are often fed too much and can't get enough exercise to keep from getting fat (Figure 9).

If you don't want your horse to get laminitis, don't let it get fat. If it does get fat, reduce the feed, use a feed with lower calorie content (for example, grass hay instead of alfalfa), increase its exercise, or all three. And, for goodness sake, be patient. You want weight loss to be gradual and consistent, as it does take time for a horse to lose weight. Ask your veterinarian for help in coming up with a safe weight-loss program.

Breed

While there's no particular breed that is genetically programmed to get laminitis more often than another, certain breeds do seem to be more susceptible to the condition. For example, Morgan horses and donkeys seem to have a tendency to get fat, and you need to watch their weight carefully.

You also have to be extra careful with ponies, which seem to be especially prone to becoming overweight. This may be because they came from areas where the natural diet was fairly poor in both quantity and quality, so perhaps ponies became "programmed" to easily store fat to get them through lean times. Most people also don't really have a good idea of how much ponies actually weigh, so it's pretty easy to overestimate how much feed is appropriate for a particular individual.

FIGURE 9

Don't overfeed your horse.

Thoroughbreds pose their own particular problem. While they aren't at risk any more than any other horse, they may tend to be flat-footed and thin-soled, which makes their feet more prone to trauma-related laminitis.

Age

Horses over 15 years are at greater risk for laminitis, most likely because, among other reasons, their activity level often gets reduced. In addition, older horses are the ones that typically develop growths on their pituitary glands (Cushing's Syndrome, see p. 44). Laminitis can be particularly hard to treat in older horses because there's no cure for old age, nor is there a cure for many of the underlying conditions that they develop. On the other hand, laminitis is rare in horses under two years of age. One exception to this general rule is seen with young horses that are being prepared for halter classes or yearling sales. These portly little fellows often get a lot of grain pushed to them and may be prone to laminitis due to the combination of obesity, grain overload and (in many cases) lack of exercise. Even if these "halter specials" don't develop laminitis, they may be more prone to develop obesity and related problems later in life.

Exercise

Regular exercise is absolutely critical. Some veterinarians maintain that fit horses rarely develop laminitis. Exercise is not only helpful in keeping bones, joints, and muscles in good working order, but it also helps your horse keep its weight under control. Whether exercise comes in the form of daily turnout or some sort of regularly structured exercise program, you need to get your horse out and moving. If you expect it to exercise hard, it's important to develop a training and conditioning program that's appropriate for the horse's work.

The *regular* part of "regular exercise" is particularly important. If you take your horse out and run it into the ground over the weekend, and don't exercise it during the week, you're going to *increase* the chance that it will develop laminitis. Inadequate conditioning can also be a problem when the weather is hot and humid or if the work is particularly hard. If you work your horse to exhaustion, imbalances in body fluids, salts, and metabolism may occur. The muscle and gastrointestinal problems that accompany such imbalances increase the risk that the horse will develop laminitis.

You also have to mind the surface that you're riding on. Hard surfaces, such as asphalt or hard-packed dirt, can cause increased pounding on the horse's feet and result in trauma-induced laminitis (road founder). This would seem to be common sense—you wouldn't want someone to make you run up and down on the pavement for endless hours, would you?

Keep in mind that this chapter is about *preventing* laminitis. If your horse develops laminitis, exercise can be harmful, particularly in the early stages. Exercising a horse with laminitis runs the risk of adding stress to the already compromised structures, and causing further damage to the foot.

Supplements

There is no known supplement that, when added to the feed, can help prevent (or treat) laminitis. Nevertheless, they seem to be everywhere, so it's useful to know about them. Supplements that have been suggested include:

1. *Magnesium* for weight reduction in overweight, cresty-necked, horses. This is based on the idea that magnesium-deficient diets and low total body stores of magnesium are associated with resistance to the hormone insulin in people. Insulin resistance appears to also be a problem in horses with obesity associated laminitis. Still, magnesium deficiency is not something that's ever been shown in the horse. Furthermore, while it is at least *possible* that some hays and grasses are low in magnesium, most are not (if you're worried, you can get a feed analysis done on your horse's hay).

   Still, there's no evidence that magnesium supplementation is effective at either preventing or treating laminitis, nor is there any evidence that it helps with weight reduction. In addition, most readily available magnesium supplements appear to be very poorly absorbed. Giving magnesium is not necessarily without problems of its own. Poisoning from oral magnesium supplementation has been reported in cattle and there's an association between high magnesium levels in the diet and/or water and the formation of intestinal "stones" (enteroliths) in horses.

2. *Fatty acid supplementation* was reported to prevent laminitis in a carbohydrate-overload model in one study a few years back. The idea behind giving fatty acids is based on the fact that because it's thought that they may help reduce hypertension in people, and there's hypertension associated with the onset of laminitis in horses, giving fatty acids might help reduce hypertension in horses, too. While giving fatty acids to horses would be extremely unlikely to do any harm (in fact, giving fat to horses in their diet has a lot going for it), there's been nothing else since the initial study to support the idea that fatty acids are useful in preventing laminitis.

3. *Hoof supplements*, such as biotin or methionine (and innumerable others), to provide "nutritional support." While there certainly appears to be little harm in such practices, and there's even some support for the idea that supplemental biotin can help some horses with poor quality hooves, there's also no evidence that any of them can help prevent laminitis.

DISEASES

As you know, laminitis is associated with many diseases, from intestinal disorders to problems with the lungs, from reproductive disorders to orthopedic conditions. When there is an underlying disease process, the foot is just an innocent bystander that gets caught up in the problem. However, if laminitis develops in a horse that's otherwise seriously ill, the laminitis can get very bad and may be very difficult to treat. Even if the horse gets over its primary problem, the secondary laminitis can be devastating, even fatal. Wouldn't it be nice if you could figure out a way to keep your horse from ever getting seriously ill?

However, what you can do is keep your veterinarian involved in your horse's health care. Make sure that your horse is seen at least once a year for a routine check-up. This is especially important in horses more than 15 years of age, those more likely to develop equine Cushing's Syndrome. Of course, if your horse gets sick, consult with your veterinarian as soon as possible, and in particular under any of the following circumstances:

- It has got into the feed bin and eaten several pounds of grain.

- It is showing signs of colic, such as pain, rolling, pawing, biting at the flanks, etc.

- It seems reluctant to move or has any of the signs of acute laminitis that were discussed earlier on p. 26.

- Your mare has not passed her afterbirth within three hours after foaling.

By making sure that your horse is healthy and by treating problems as soon as they come up, you'll go a long way toward preventing more serious problems.

MEDICATIONS
It's not a good idea to routinely give your horse medication in the absence of veterinary supervision. Certainly, not all drugs are harmful, however, all effective medications have potential side effects of which you may not be aware. Be particularly careful with the use of corticosteroid drugs.

TRANSPORT
In this day of increased mobility, horses are moving around a lot more. Fortunately, serious shipping problems are not particularly common, but transporting horses does increase the risk for a number of medical problems, including laminitis. Procedures and rules to follow that may decrease the chance of shipping-related problems include:

1. Ensure that the horse's vaccinations are current.

2. Don't ship if your horse is sick.

3. Don't make any dramatic changes prior to shipping. For example, don't feed new hay or don't deworm your horse the morning before your trip.

4. Do not ship in very hot weather.

5. Ensure that the horse stays well hydrated during the trip.

6. No grain while shipping—the horse can't exercise or move around to help work off the extra calories.

7. Provide adequate ventilation to help decrease respiratory problems.

8. Wet the hay to keep the dust down.

9. Consider stopping and allowing the horse to move around and lower its head every 8 hours. Keeping a horse's head tied up during transport prevents gravity from helping to clear out its breathing passages, and the risk of infection involving the lungs and/or chest cavity goes up, accordingly. Laminitis is a common complication of such infections.

After you get to your destination, you'll want to get the horse out and exercising to get the "kinks" out. Also, make sure that it's eating and drinking and keep track of its temperature for a few days. If the horse doesn't act normally, or if it starts to develop a fever (normal temperature is generally between 99 and 101 degrees Fahrenheit) you should call your veterinarian immediately.

FOOT CARE

The old adage "No foot, no horse" is certainly true. Routine care of and regular attention to your horse's feet is essential. You should clean and inspect your horse's feet every day and make sure that problems such as thrush, cracks, and separations are attended to. Your horse should be shod or trimmed on a regular schedule, determined with the advice of your veterinarian and/or farrier. If your horse throws a shoe, or steps on a nail, get it seen to as soon as possible, and protect the foot in the interim with a boot or bandage.

Virginiamycin

In some parts of the world (not the U.S.), a product known as *Founderguard®* is available. It's actually an antibiotic called *virginiamycin*. Studies have shown that feeding this antibiotic regularly at low levels does help prevent laminitis in horses that are kept on pasture. Unfortunately, it has little value as a *treatment* for laminitis.

Even though virginiamycin does appear to help prevent laminitis, some people have concerns about feeding antibiotics to animals. It's not that it hurts the animal, rather, it most likely tends to add to the very real problem of antibiotic resistance. When bacteria are exposed to antibiotics, over time, they can build up a resistance to them. As more and more antibiotic-resistant bacteria are selected, infections—*any* infection—become more difficult to treat. Feeding antibiotics on a routine basis is not a particularly good idea and certainly should not be substituted for good management practices.

Not all cases of laminitis can be prevented. However, you've got your best chance at getting your horse back to normal if you take proper care of it and recognize any problems as early as possible. Remember, an episode of laminitis can affect your horse for the rest of its life. Accordingly, a good prevention program should be a life-long task.

Dr. Chris King is a 1985 graduate of the University of Queensland (Australia). She spent several years in equine practice, then followed with a residency in equine medicine and surgery and a Masters degree in equine exercise physiology. She moved to the U.S. in 1993 to work at North Carolina State University. She left NCSU and now works as a veterinary medical writer. Dr. King has also co-authored an excellent book on laminitis, Preventing Laminitis in Horses, *Paper Horse Publishing, Cary, North Carolina, 2000. It's a wonderful source of information.*

A Philosophy of Caring for the Horse with Laminitis

with William Moyer, DVM

As you've seen, there are any number of techniques and approaches to caring for the horse with laminitis, ranging from the simple—controlling pain; some trimming and shoeing—to the complex. Indeed, one of the greatest challenges is trying to choose from all of the different treatment options. That's mostly because there is no single one that has been shown to be effective in all—or even most—cases. Furthermore, studies to compare the usefulness of the various approaches have simply not been done. Thus, decisions concerning treatment provided by a veterinarian, farrier, or another involved party (such as an insurance company) are going to be invariably based on that particular party's experience.

Most of the techniques described in the veterinary medical literature for the treatment of horses with laminitis probably have some use in some circumstances. However, that being said, no particular treatment has ever been shown to be better than another. Indeed, because each case of laminitis can be somewhat different, the "ideal" treatment, even if it were known, would probably be different from one horse to another. That means that for your particular situation, some trial and error may be inevitable.

Not only are there many different treatment approaches, but also the ability of a particular individual to use those approaches may vary. Not every veterinarian is comfortable performing unusual surgeries, and not every farrier has made every possible shoe. Complicated and difficult cases may require some considerable teamwork, and it will be up to you to keep everyone working together. But, caring for a horse with laminitis is more than just finding the best therapeutic approach, you need to consider some other important facts.

First, the foot of the horse is attached to an entire horse. As the owner of that horse, you need to understand that there are many other problems that can be associated with the condition. Laminitis may be associated with any number of disease entities and for treatment to be successful the primary problem must be addressed (if possible).

Second, you also have to keep the horse's well being in mind. Most horses that have laminitis also have a significant amount of pain, lameness, and/or structural change within the hoof itself. So much damage may occur that healing of the foot is impossible. So, depending on what happens, and regardless of what you do and how well you do it, you may not have a successful outcome, no matter how early in the course of the disease you start treatment.

Third, you also have to realize that no one can really predict how an individual horse with laminitis is going to respond to treatment. Horses that contract the condition will have a wide variation in the signs of disease, as well as in the extent of the underlying problem. The signs of disease and the amount of pain do not always correspond to the degree of damage in the foot. Thus, trying to predict how things are going to go may be an exercise in futility. You have to try to keep your expectations realistic.

Realize that it may not be possible to determine the exact cause of the problem and that it's not always someone's fault. Sure, in some circumstances you are going to know that your horse broke into the grain bin, or that it has been really sick, but in other cases, the horse will have been normal and there will have been no obvious changes in his regimen. Or, while laminitis may occasionally follow the administration of some drugs, administration of those drugs may not necessarily be the sole cause of the horse's laminitis. For example, even though laminitis may follow the administration of corticosteroid drugs, nobody seems to be able to cause laminitis in healthy horses simply by giving them corticosteroids. If you start speculating as to what caused the problem, you may find out that all you end up doing is placing blame somewhere that it doesn't belong.

You must realize that the mechanism and development of laminitis has not been fully worked out. This means that there are a lot of theories, and some facts, but as yet an incomplete understanding of the condition. Accordingly, don't accept speculation as truth. No one has all the answers.

Understand that by the time laminitis is recognized (that is, the horse is in pain or showing lameness) structural changes within the foot have already occurred. Thus, anyone who tries to get in and treat the problem is already "behind the eight ball," since the disease got a head start. At that point, treatment becomes a matter of trying to institute damage control, and success largely depends on the extent of it. The time frame may be easily lengthened if the horse in question, for whatever reason, has not been observed carefully. For example, there are some horses—say, those kept out in a pasture—that may become afflicted with laminitis, but not noticed. In those horses in which the problem has been ongoing, the damage to the laminae and the hoof may be so extensive that it may not be possible to fix them.

Know that no matter how good or bad things look at the start, you can't necessarily predict how things are going to turn out. You may fix things, and get a sound riding horse back, but you could also end up with a permanently lame horse, or even one that must be put to sleep. The best that you can do is attack the condition aggressively and sensibly, and hope for a successful outcome.

It's also possible that you won't end up with a *complete* cure. Even if the initial treatment stops the progress of the disease (or it stops on its own), you still might end up with a horse that requires special attention. It may need regular visits from your veterinarian and/or farrier and special medications or horseshoes. There may be special feed or bedding requirements. You could have a life-long problem on your hands. The best that you may be able to do is try to figure out how the horse can get along with permanent damage to its feet. Even if the horse is able to resume its athletic career, its feet may never again be normal. Needless to say, there's a cost for this—managing laminitis can be very expensive.

Recognize that even if things go well initially, other foot problems may crop up later. These problems include not only additional episodes of laminitis, but also such things as foot abscesses, separations of the hoof wall ("seedy toe"), abnormal hoof growth or lameness. Even if a horse appears to have fully recovered, it's still at risk of developing the condition again at some point in the future.

It's probably not a bad idea to read over these pages several times when treating a horse with laminitis. The more you know, the less likely you will be surprised by a particular outcome. This is a complicated disease and one that can be very difficult to manage.

One of the most difficult issues for you to consider is whether or not to keep a horse alive that fails to respond to any treatment and is in chronic pain as a result. It's a humane issue. Unfortunately, there's rarely an end point where you can just say, "Now's the time." It seems as if there's always a bit of hope, even in the worst of cases. It's not like dealing with a horse that has a severely broken leg, a tragic circumstance to be sure, but one for which the options are usually pretty clear. Instead, you may need to take a look at a horse with severe and non-responsive laminitis, and ask yourself, "Is this horse being kept alive for its sake or for my sake?" Clearly, the decision to end a horse's life is not one that should be made easily, and it's a decision that should be made by all of the parties involved in the care of the horse. But, keeping a chronically suffering horse alive just because someone thinks that the horse is irreplaceable is not kind—it's cruel.

Finally, if someone tells you that if he or she had started a particular therapy "in time," your horse would be much better, you're either dealing with a charlatan or an egomaniac. There is absolutely no way that anyone could support such a statement. Laminitis is a humbling disease and anyone who claims universal success simply hasn't treated enough horses.

Horses with laminitis can be difficult to take care of even under the best of circumstances. Unfortunately, at this time, there is no single "optimum" therapy that can benefit each horse. It seems as if most of the things that are done to treat horses with laminitis have *some* value, at least some of the time, but no one procedure is clearly better than another. Learning as much as you can about this condition is very important for a successful outcome, and, failing that, for understanding that you did everything that you could.

Dr. Bill Moyer is a 1970 graduate of Colorado State University. After completing an internship and residency in equine medicine and surgery at the University of Pennsylvania, he practiced in that area until 1980. Returning to the University of Pennsylvania, he stayed for 13 years, becoming a professor of Sports Medicine. In 1993, Dr. Moyer became Professor and Head of the Department of Large Animal Medicine at Texas A&M University, a position he currently holds. Dr. Moyer's professional interests include equine lameness, shoeing and foot problems, and racetrack surface studies. He's a legend in the horse world.

BIBLIOGRAPHY

Treatment

Hoff T.K., D.M. Hood, and I.P.Wagner. "Effectiveness of Glyceryl Trinitrate for Enhancing Digital Submural Perfusion in Horses." *Am J Vet Res.* (2002): 63(5), 648-52.

Erkert R.S., and C.G. Macallister. "Isoxsuprine Hydrochloride in the Horse: A Review." *J Vet Pharmacol Ther.* (2002): 25(2), 81-7.

Brumbaugh G.W., Lopez H. Sumano, and, Sepulveda M.L. Hoyas. "The Pharmacologic Basis for the Treatment of Developmental and Acute Laminitis." *Veterinary Clinics of North America Equine Practice* (1999): 15(2), 345-62.

Parks A.H., O.K. Balch, and M.A. Collier. "Treatment of Acute Laminitis: Supportive Therapy." *Veterinary Clinics of North America Equine Practice* (1999): 15(2), 363-74.

Ingle-Fehr J.E., and G.M. Baxter. "The Effect of Oral Isoxsuprine and Pentoxifylline on Digital and Laminar Blood Flow in Healthy Horses." *Vet Surg.* (1999): 28(3), 154-60.

Owens J.G., S.G. Kamerling, and S.R. Stanton, et al. "Effects of Ketoprofen and Phenylbutazone on Chronic Hoof Pain and Lameness in the Horse." *Equine Vet J.* (1995): 27(4), 296-300.

Risk Factors

Johnson, P.J. "The Equine Metabolic Syndrome (Peripheral Cushing's Syndrome)." *Veterinary Clinics of North America Equine Practice* (2002): 18, 271-293.

Johnson, P.J., S.H. Slight, Ganjam, V. K., et al. "Glucocorticoids and Laminitis in the Horse." *Veterinary Clinics of North America Equine Practice* (2002): 18, 219-236.

Alford P., S. Geller, B. Richrdson, et al. "A Multicenter, Matched Case-Control Study of Risk Factors for Equine Laminitis." *Prev Vet Med.* (2001): 49(3-4), 209-22.

Hoffman R.M., J.A.Wilson, D.S. Kronfeld, et al. "Hydrolyzable Carbohydrates in Pasture, Hay, and Horse Feeds: Direct Assay and Seasonal Variation." *Journal Animal Science* (2001): 79(2), 500-6.

French K., C.C. Pollitt, and M.A. Pass. "Pharmacokinetics and Metabolic Effects of Triamcinolone Acetonide and Their Possible Relationships to Glucocorticoid-Induced Laminitis in Horses." *J Vet Pharmacol Ther.* (2000): 23(5), 287-92.

Rohrbach B.W., E.M.Green, J.W. Oliver, et al. "Aggregate Risk Study of Exposure to Endophyte-Infected (Acremonium coenophialum) Tall Fescue as a Risk Factor for Laminitis in Horses." *Am J Vet Res.* (1995): 56(1), 22-6.

Polzer J. and M.R. Slater. "Age, Breed, Sex and Seasonality as Risk Factors for Equine Laminitis." *Prev Vet Med.* (1997): 29(3), 179-84.

Development

Mungall B.A., M. Kyaw-Tanner, and C.C. Pollitt. "In Vitro Evidence for a Bacterial Pathogenesis of Equine Laminitis." *Vet Microbiol.* (2001): 79(3), 209-23.

Adair H.S. 3rd, D.O. Goble, J.L. Schmidhammer, et al. "Laminar Microvascular Flow, Measured by Means of Laser Doppler Flowmetry, During the Prodromal Stages of Black Walnut-Induced Laminitis in Horses." *Am J Vet Res.* (2000): 61(8), 862-8.

Hood D.M., I.P. Wagner, and G.W. Brumbaugh. "Evaluation of Hoof Wall Surface Temperature as an Index of Digital Vascular Perfusion During the Prodromal and Acute Phases of Carbohydrate-Induced Laminitis in Horses." *Am J Vet Res.* (2001): 62(7), 1167-72.

Weiss D.J., O.A. Evanson, B.T. Green, et al. "In Vitro Evaluation of Intraluminal Factors that may Alter Intestinal Permeability in Ponies with Carbohydrate-Induced Laminitis." *Am J Vet Res.* (2000): 61(8), 858-61.

Pollitt, C.C. "Equine Laminitis: A Revised Pathophysiology." *Proc 45th Ann Am Assn Eq Prac.* (1999): 188-192.

Pollitt C.C. and C.T. Davies. "Equine Laminitis: Its Development Coincides with Increased Sublamellar Blood Flow." *Equine Vet J Suppl.* (1998): (26), 125-32.

Pollitt C.C. "Basement Membrane Pathology: A Feature of Acute Equine Laminitis." *Equine Vet J.* (1996): 28(1): 38-46.

Molyneux G.S., C.J. Haller, K. Mogg, et al. "The Structure, Innervation and Location of Arteriovenous Anastomoses in the Equine Foot. *Equine Vet J.* (1994): 26(4), 305-12.

General

Mansmann, R.A. and C. King. *Preventing Laminitis in Horses—A Practical Guide to Decreasing the Risk of Laminitis (Founder) in Your Horse.* Paper Horse Publishing, 2000.

Sloet van Oldruitenborgh-Oosterbaan, M.M. "Laminitis in the Horse: A Review." *Vet Q.* (1999): 21(4), 121-7.

Hood D.M. "Laminitis as a Systemic Disease." *Vet Clin North Am Equine Pract.* (1999): 15(2), 481-94.

Cripps P.J. and R.A. Eustace. "Radiological Measurements from the Feet of Normal Horses with Relevance to Laminitis." *Equine Vet J.* (1999): 31(5), 427-32.

Hood D.M. "The Mechanisms and Consequences of Structural Failure of the Foot." *Vet Clin North Am Equine Pract.* (1999): 15(2), 437-61.

Grosenbaugh D.A., S.J. Morgan SJ, and D.M. Hood. "The Digital Pathologies of Chronic Laminitis." *Vet Clin North Am Equine Pract.* (1999): 15(2), 419-36.

Morgan S.J., D.A. Grosenbaugh, and D.M. Hood. "The Pathophysiology of Chronic Laminitis. Pain and Anatomic Pathology." *Vet Clin North Am Equine Pract.* (1999): 15(2), 395-417.

Herthel D. and D.M. Hood. "Clinical Presentation, Diagnosis, and Prognosis of Chronic Laminitis." *Vet Clin North Am Equine Pract.* (1999): 15(2), 375-94.

Hood D.M. "The Pathophysiology of Developmental and Acute Laminitis." *Vet Clin North Am Equine Pract.* (1999): 15(2), 321-43.

Swanson T.D. "Clinical Presentation, Diagnosis, and Prognosis of Acute Laminitis." *Vet Clin North Am Equine Pract.* (1999): 15(2), 311-9

Hunt R.J. "A Retrospective Evaluation of Laminitis in Horses." *Equine Vet J.* (1993): 25(1), 61-4.

Surgery

Eastman T.G., C.M. Honnas, B.A. Hague, et al. "Deep Digital Flexor Tenotomy as a Treatment for Chronic Laminitis in Horses: 35 Cases (1988-1997)." *J Am Vet Med Assoc.* (1999): 214(4), 517-9.

Prognosis

Cripps P.J. and R.A. Eustace. "Factors Involved in the Prognosis of Equine Laminitis in the UK." *Equine Vet J.* (1999): 31(5), 433-42.

Stick J.A., H.W. Jann, E.A. Scott, et al. "Pedal Bone Rotation as a Prognostic Sign in Laminitis of Horses." *J Am Vet Med Assoc.* (1982): 180(3), 251-3.

INDEX

Page numbers in *italics* indicate figures.